Particle Physics & Representation Theory

Contents

Chapter 1

Particle physics and representation theory

There is a natural connection between particle physics and representation theory, as first noted in the 1930s by Eugene Wigner.[1] It links the properties of elementary particles to the structure of Lie groups and Lie algebras. According to this connection, the different quantum states of an elementary particle give rise to an irreducible representation of the Poincaré group. Moreover, the properties of the various particles, including their spectra, can be related to representations of Lie algebras, corresponding to "approximate symmetries" of the universe.

1.1 General picture

In quantum mechanics, any particular particle (with a given momentum distribution, location distribution, spin state, etc.) is written as a vector (or "ket") in a Hilbert space H. To help understand what types of particles can exist, it is important to classify the possibilities for H, and their properties. The particle is more precisely characterized by the associated *projective* Hilbert space **PH**, since two vectors that differ by a scalar factor (or in physics terminology, two "kets" that differ by a "phase factor") correspond to the same physical quantum state.

Let G be the *symmetry group of the universe* – that is, the set of symmetries under which the laws of physics are invariant. (For example, one element of G is the simultaneous translation of all particles and fields forward in time by five seconds.) Starting with a particular particle in the state ket $|p_0\rangle$, and a symmetry transformation g in G, it is possible to apply the symmetry transformation to the particle to get a new state ket $|p_g\rangle = g|p_0\rangle$. For this picture to be consistent, it is necessary that **PH** holds a projective group representation of G. (For example, this condition guarantees that applying a symmetry transformation, then applying its inverse transformation, will restore the original quantum state.)

Therefore, any given particle is associated with a unique (possibly trivial) representation of G on a projective vector space **PH**. (We say the particle "lies in", or "transforms as" the representation.) In many important cases, it can be shown that the particle is also (more specifically) associated with a group representation of G on the underlying (non-projective) space H.[2] Wigner's Theorem proves that it is a unitary representation, or possibly anti-unitary.[2]

So we conclude that each type of particle corresponds to a representation of G, and if we can classify the group representations of G, we will have much more information about the possibilities and properties of H, and hence what types of particles can exist.

1.2 Poincaré group

The group of translations and Lorentz transformations form the Poincaré group, and this group is certainly a subgroup of G (neglecting general relativity effects, or in other words, in flat space). Hence, any representation of G will in particular be a representation of the Poincaré group. Representations of the Poincaré group are in many cases characterized by a nonnegative mass and a half-integer spin (see Wigner's classification); this can be thought of as the reason that particles have quantized spin. (Note that there are in fact other possible representations, such as tachyons, infraparticles, etc., which in some cases do not have quantized spin or fixed mass.)

1.3   Other symmetries

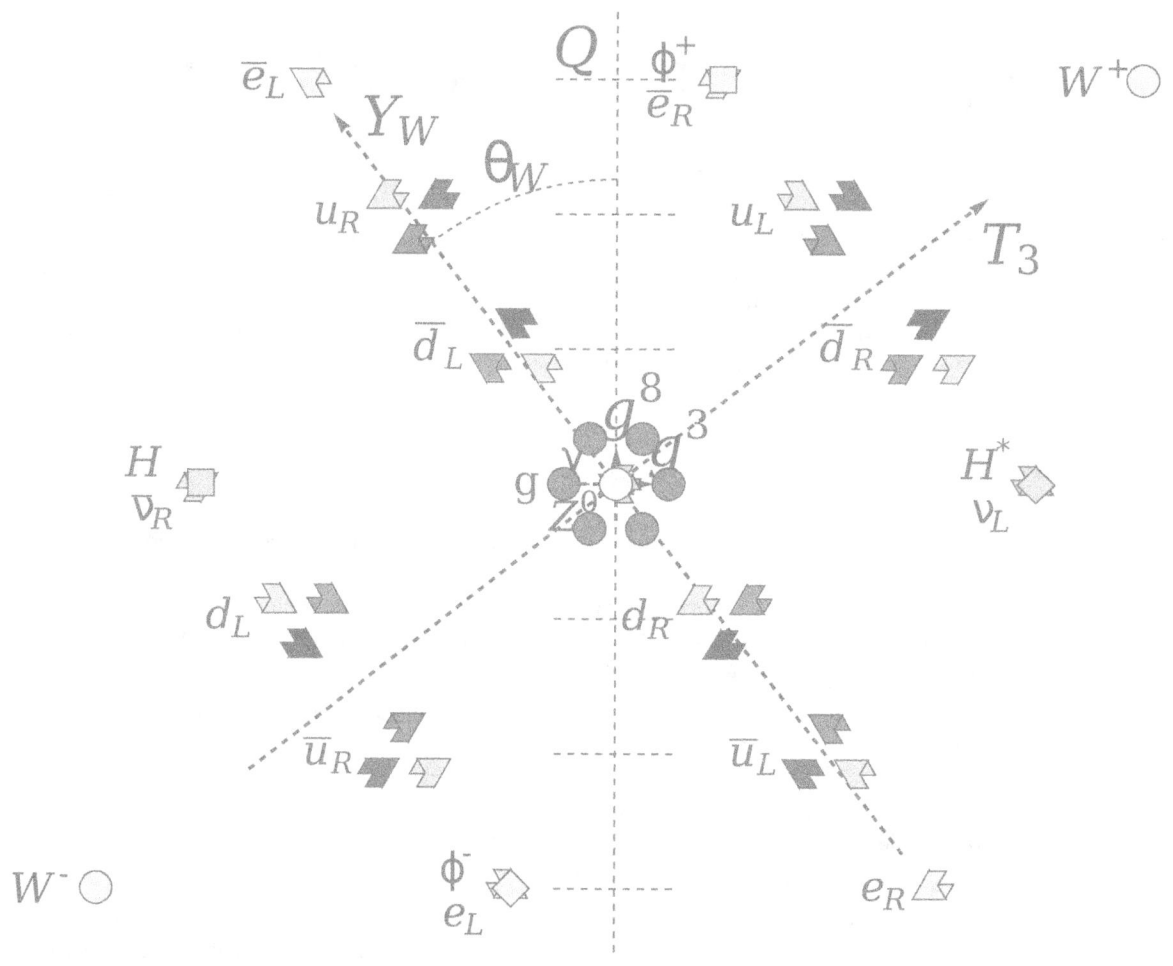

The pattern of weak isospins, weak hypercharges, and color charges (weights) of all known elementary particles in the Standard Model, rotated by the weak mixing angle to show electric charge roughly along the vertical.

While the spacetime symmetries in the Poincaré group are particularly easy to visualize and believe, there are also other types of symmetries, called internal symmetries. One example is color SU(3), an exact symmetry corresponding to the continuous interchange of the three quark colors.

1.4   Approximate symmetries

Although the above symmetries are believed to be exact, other symmetries are only approximate.

1.4.1 Hypothetical example

As an example of what an approximate symmetry means, suppose we lived inside an infinite ferromagnet, with magnetization in some particular direction. An experimentalist in this situation would find not one but two distinct types of electrons: one with spin along the direction of the magnetization, with a slightly lower energy (and consequently, a lower mass), and one with spin anti-aligned, with a higher mass. Our usual SO(3) rotational symmetry, which ordinarily connects the spin-up electron with the spin-down electron, has in this hypothetical case become only an *approximate* symmetry, relating *different types of particles* to each other.

1.4.2 Lie algebras versus Lie groups

Many (but not all) symmetries or approximate symmetries, for example the ones above, form Lie groups. Rather than study the representation theory of these Lie groups, it is often preferable to study the closely related representation theory of the corresponding Lie algebras, which are usually simpler to compute.

1.4.3 General definition

In general, an approximate symmetry arises when there are very strong interactions that obey that symmetry, along with weaker interactions that do not. In the electron example above, the two "types" of electrons behave identically under the strong and weak forces, but differently under the electromagnetic force.

1.4.4 Example: isospin symmetry

Main article: Isospin

An example from the real world is isospin symmetry, an SU(2) group corresponding to the similarity between up quarks and down quarks. This is an approximate symmetry: While up and down quarks are identical in how they interact under the strong force, they have different masses and different electroweak interactions. Mathematically, there is an abstract two-dimensional vector space

$$\text{quark up} \to \begin{pmatrix} 1 \\ 0 \end{pmatrix}, \qquad \text{quark down} \to \begin{pmatrix} 0 \\ 1 \end{pmatrix}$$

and the laws of physics are *approximately* invariant under applying a determinant-1 unitary transformation to this space:[3]

$$\begin{pmatrix} x \\ y \end{pmatrix} \mapsto A \begin{pmatrix} x \\ y \end{pmatrix}, \quad \text{where } A \text{ in is } SU(2)$$

For example, $A = \begin{pmatrix} 0 & 1 \\ -1 & 0 \end{pmatrix}$ would turn all up quarks in the universe into down quarks and vice versa. Some examples help clarify the possible effects of these transformations:

- When these unitary transformations are applied to a proton, it can be transformed into a neutron, or into a superposition of a proton and neutron, but not into any other particles. Therefore, the transformations move the proton around a two-dimensional space of quantum states. The proton and neutron are called an "isospin doublet", mathematically analogous to how a spin-½ particle behaves under ordinary rotation.

- When these unitary transformations are applied to any of the three pions ($\pi 0$, $\pi +$, and $\pi -$), it can change any of the pions into any other, but not into any non-pion particle. Therefore, the transformations move the pions around a three-dimensional space of quantum states. The pions are called an "isospin triplet", mathematically analogous to how a spin-1 particle behaves under ordinary rotation.

- These transformations have no effect at all on an electron, because it contains neither up nor down quarks. The electron is called an isospin singlet, mathematically analogous to how a spin-0 particle behaves under ordinary rotation.

In general, particles form isospin multiplets, which correspond to irreducible representations of the Lie algebra SU(2). Particles in an isospin multiplet have very similar but not identical masses, because the up and down quarks are very similar but not identical.

1.4.5   Example: flavour symmetry

Isospin symmetry can be generalized to flavour symmetry, an SU(3) group corresponding to the similarity between up quarks, down quarks, and strange quarks.[3] This is, again, an approximate symmetry, violated by quark mass differences and electroweak interactions—in fact, it is a poorer approximation than isospin, because of the strange quark's noticeably higher mass.

Nevertheless, particles can indeed be neatly divided into groups that form irreducible representations of the Lie algebra SU(3), as first noted by Murray Gell-Mann and independently by Yuval Ne'eman (see the eightfold way).

1.5   See also

- Charge (physics)

- Lie algebra

- Lie group

- Poincaré group

- Representation theory:

  - Of Lie algebras

  - Of Lie groups

  - Representation theory of the Poincaré group

- Special unitary group

1.6   Notes

[1] Wigner received the Nobel Prize in Physics in 1963 "for his contributions to the theory of the atomic nucleus and the elementary particles, particularly through the discovery and application of fundamental symmetry principles"; see also Wigner's theorem, Wigner's classification.

[2] See Weinberg (1995), Chapter 2 appendix A and B.

[3] Lecture notes by Prof. Mark Thomson

1.7   References

- Coleman, Sidney (1985) *Aspects of Symmetry: Selected Erice Lectures of Sidney Coleman.* Cambridge Univ. Press. ISBN 0-521-26706-4.

- Georgi, Howard (1999) *Lie Algebras in Particle Physics.* Reading, Massachusetts: Perseus Books. ISBN 0-7382-0233-9.

- Hall, Brian C., (2006) *Lie Groups, Lie Algebras, and Representations: An Elementary Introduction.* Springer. ISBN 0-387-40122-9.

- Sternberg, Shlomo (1994) *Group Theory and Physics.* Cambridge Univ. Press. ISBN 0-521-24870-1. Especially pp. 148–150.

- Steven Weinberg (1995). *The Quantum Theory of Fields, Volume 1: Foundations.* Cambridge Univ. Press. ISBN 0-521-55001-7. Especially appendices A and B to Chapter 2.

1.8 External links

- Baez, John C.; Huerta, John (2009). "The Algebra of Grand Unified Theories". arXiv:0904.1556.

Chapter 2

Group theory

This article covers advanced notions. For basic topics, see Group (mathematics).
For group theory in social sciences, see social group.

In mathematics and abstract algebra, **group theory** studies the algebraic structures known as groups. The concept of a group is central to abstract algebra: other well-known algebraic structures, such as rings, fields, and vector spaces, can all be seen as groups endowed with additional operations and axioms. Groups recur throughout mathematics, and the methods of group theory have influenced many parts of algebra. Linear algebraic groups and Lie groups are two branches of group theory that have experienced advances and have become subject areas in their own right.

Various physical systems, such as crystals and the hydrogen atom, may be modelled by symmetry groups. Thus group theory and the closely related representation theory have many important applications in physics, chemistry, and materials science. Group theory is also central to public key cryptography.

One of the most important mathematical achievements of the 20th century[1] was the collaborative effort, taking up more than 10,000 journal pages and mostly published between 1960 and 1980, that culminated in a complete classification of finite simple groups.

2.1 Main classes of groups

Main articles: Group (mathematics) and Glossary of group theory

The range of groups being considered has gradually expanded from finite permutation groups and special examples of matrix groups to abstract groups that may be specified through a presentation by generators and relations.

2.1.1 Permutation groups

The first class of groups to undergo a systematic study was permutation groups. Given any set X and a collection G of bijections of X into itself (known as *permutations*) that is closed under compositions and inverses, G is a group acting on X. If X consists of n elements and G consists of *all* permutations, G is the symmetric group Sn; in general, any permutation group G is a subgroup of the symmetric group of X. An early construction due to Cayley exhibited any group as a permutation group, acting on itself ($X = G$) by means of the left regular representation.

In many cases, the structure of a permutation group can be studied using the properties of its action on the corresponding set. For example, in this way one proves that for $n \geq 5$, the alternating group An is simple, i.e. does not admit any proper normal subgroups. This fact plays a key role in the impossibility of solving a general algebraic equation of degree $n' \geq 5$ *in radicals*.

The popular puzzle Rubik's cube invented in 1974 by Ernő Rubik has been used as an illustration of permutation groups.

2.1.2 Matrix groups

The next important class of groups is given by *matrix groups*, or linear groups. Here G is a set consisting of invertible matrices of given order n over a field K that is closed under the products and inverses. Such a group acts on the n-dimensional vector space K^n by linear transformations. This action makes matrix groups conceptually similar to permutation groups, and the geometry of the action may be usefully exploited to establish properties of the group G.

2.1.3 Transformation groups

Permutation groups and matrix groups are special cases of transformation groups: groups that act on a certain space X preserving its inherent structure. In the case of permutation groups, X is a set; for matrix groups, X is a vector space.

The concept of a transformation group is closely related with the concept of a symmetry group: transformation groups frequently consist of *all* transformations that preserve a certain structure.

The theory of transformation groups forms a bridge connecting group theory with differential geometry. A long line of research, originating with Lie and Klein, considers group actions on manifolds by homeomorphisms or diffeomorphisms. The groups themselves may be discrete or continuous.

2.1.4   Abstract groups

Most groups considered in the first stage of the development of group theory were "concrete", having been realized through numbers, permutations, or matrices. It was not until the late nineteenth century that the idea of an abstract group as a set with operations satisfying a certain system of axioms began to take hold. A typical way of specifying an abstract group is through a presentation by *generators and relations*,

$$G = \langle S | R \rangle.$$

A significant source of abstract groups is given by the construction of a *factor group*, or quotient group, G/H, of a group G by a normal subgroup H. Class groups of algebraic number fields were among the earliest examples of factor groups, of much interest in number theory. If a group G is a permutation group on a set X, the factor group G/H is no longer acting on X; but the idea of an abstract group permits one not to worry about this discrepancy.

The change of perspective from concrete to abstract groups makes it natural to consider properties of groups that are independent of a particular realization, or in modern language, invariant under isomorphism, as well as the classes of group with a given such property: finite groups, periodic groups, simple groups, solvable groups, and so on. Rather than exploring properties of an individual group, one seeks to establish results that apply to a whole class of groups. The new paradigm was of paramount importance for the development of mathematics: it foreshadowed the creation of abstract algebra in the works of Hilbert, Emil Artin, Emmy Noether, and mathematicians of their school.

2.1.5   Topological and algebraic groups

An important elaboration of the concept of a group occurs if G is endowed with additional structure, notably, of a topological space, differentiable manifold, or algebraic variety. If the group operations m (multiplication) and i (inversion),

$$m : G \times G \to G, (g, h) \mapsto gh, \quad i : G \to G, g \mapsto g^{-1},$$

are compatible with this structure, i.e. are continuous, smooth or regular (in the sense of algebraic geometry) maps, then G becomes a topological group, a Lie group, or an algebraic group.[2]

The presence of extra structure relates these types of groups with other mathematical disciplines and means that more tools are available in their study. Topological groups form a natural domain for abstract harmonic analysis, whereas Lie groups (frequently realized as transformation groups) are the mainstays of differential geometry and unitary representation theory. Certain classification questions that cannot be solved in general can be approached and resolved for special subclasses of groups. Thus, compact connected Lie groups have been completely classified. There is a fruitful relation between infinite abstract groups and topological groups: whenever a group Γ can be realized as a lattice in a topological group G, the geometry and analysis pertaining to G yield important results about Γ. A comparatively recent trend in the theory of finite groups exploits their connections with compact topological groups (profinite groups): for example, a single p-adic analytic group G has a family of quotients which are finite p-groups of various orders, and properties of G translate into the properties of its finite quotients.

2.2   Branches of group theory

2.2.1 Finite group theory

Main article: Finite group

During the twentieth century, mathematicians investigated some aspects of the theory of finite groups in great depth, especially the local theory of finite groups and the theory of solvable and nilpotent groups. As a consequence, the complete classification of finite simple groups was achieved, meaning that all those simple groups from which all finite groups can be built are now known.

During the second half of the twentieth century, mathematicians such as Chevalley and Steinberg also increased our understanding of finite analogs of classical groups, and other related groups. One such family of groups is the family of general linear groups over finite fields. Finite groups often occur when considering symmetry of mathematical or physical objects, when those objects admit just a finite number of structure-preserving transformations. The theory of Lie groups, which may be viewed as dealing with "continuous symmetry", is strongly influenced by the associated Weyl groups. These are finite groups generated by reflections which act on a finite-dimensional Euclidean space. The properties of finite groups can thus play a role in subjects such as theoretical physics and chemistry.

2.2.2 Representation of groups

Main article: Representation theory

Saying that a group G *acts* on a set X means that every element of G defines a bijective map on the set X in a way compatible with the group structure. When X has more structure, it is useful to restrict this notion further: a representation of G on a vector space V is a group homomorphism:

$$\varrho : G \to \mathrm{GL}(V),$$

where $\mathrm{GL}(V)$ consists of the invertible linear transformations of V. In other words, to every group element g is assigned an automorphism $\varrho(g)$ such that $\varrho(g) \circ \varrho(h) = \varrho(gh)$ for any h in G.

This definition can be understood in two directions, both of which give rise to whole new domains of mathematics.[3] On the one hand, it may yield new information about the group G: often, the group operation in G is abstractly given, but via ϱ, it corresponds to the multiplication of matrices, which is very explicit.[4] On the other hand, given a well-understood group acting on a complicated object, this simplifies the study of the object in question. For example, if G is finite, it is known that V above decomposes into irreducible parts. These parts in turn are much more easily manageable than the whole V (via Schur's lemma).

Given a group G, representation theory then asks what representations of G exist. There are several settings, and the employed methods and obtained results are rather different in every case: representation theory of finite groups and representations of Lie groups are two main subdomains of the theory. The totality of representations is governed by the group's characters. For example, Fourier polynomials can be interpreted as the characters of U(1), the group of complex numbers of absolute value *1*, acting on the L^2-space of periodic functions.

2.2.3 Lie theory

Main article: Lie group

A Lie group is a group that is also a differentiable manifold, with the property that the group operations are compatible with the smooth structure. Lie groups are named after Sophus Lie, who laid the foundations of the theory of continuous transformation groups. The term *groupes de Lie* first appeared in French in 1893 in the thesis of Lie's student Arthur Tresse, page 3.[5]

Lie groups represent the best-developed theory of continuous symmetry of mathematical objects and structures, which makes them indispensable tools for many parts of contemporary mathematics, as well as for modern theoretical physics.

They provide a natural framework for analysing the continuous symmetries of differential equations (differential Galois theory), in much the same way as permutation groups are used in Galois theory for analysing the discrete symmetries of algebraic equations. An extension of Galois theory to the case of continuous symmetry groups was one of Lie's principal motivations.

2.2.4   Combinatorial and geometric group theory

Main article: Geometric group theory

Groups can be described in different ways. Finite groups can be described by writing down the group table consisting of all possible multiplications $g \bullet h$. A more compact way of defining a group is by *generators and relations*, also called the *presentation* of a group. Given any set F of generators $\{g_i\}i \in I$, the free group generated by F subjects onto the group G. The kernel of this map is called subgroup of relations, generated by some subset D. The presentation is usually denoted by $\langle F \mid D \rangle$. For example, the group $\mathbf{Z} = \langle a \mid \rangle$ can be generated by one element a (equal to +1 or −1) and no relations, because $n \cdot 1$ never equals 0 unless n is zero. A string consisting of generator symbols and their inverses is called a *word*.

Combinatorial group theory studies groups from the perspective of generators and relations.[6] It is particularly useful where finiteness assumptions are satisfied, for example finitely generated groups, or finitely presented groups (i.e. in addition the relations are finite). The area makes use of the connection of graphs via their fundamental groups. For example, one can show that every subgroup of a free group is free.

There are several natural questions arising from giving a group by its presentation. The *word problem* asks whether two words are effectively the same group element. By relating the problem to Turing machines, one can show that there is in general no algorithm solving this task. Another, generally harder, algorithmically insoluble problem is the group isomorphism problem, which asks whether two groups given by different presentations are actually isomorphic. For example, the additive group $\mathbf{Z}$ of integers can also be presented by

$$\langle x, y \mid xyxyx = e \rangle$$

it may not be obvious that these groups are isomorphic.[7]

Geometric group theory attacks these problems from a geometric viewpoint, either by viewing groups as geometric objects, or by finding suitable geometric objects a group acts on.[8] The first idea is made precise by means of the Cayley graph, whose vertices correspond to group elements and edges correspond to right multiplication in the group. Given two elements, one constructs the word metric given by the length of the minimal path between the elements. A theorem of Milnor and Svarc then says that given a group G acting in a reasonable manner on a metric space X, for example a compact manifold, then G is quasi-isometric (i.e. looks similar from a distance) to the space X.

2.3   Connection of groups and symmetry

Main article: Symmetry group

Given a structured object X of any sort, a symmetry is a mapping of the object onto itself which preserves the structure. This occurs in many cases, for example

1. If X is a set with no additional structure, a symmetry is a bijective map from the set to itself, giving rise to permutation groups.

2. If the object X is a set of points in the plane with its metric structure or any other metric space, a symmetry is a bijection of the set to itself which preserves the distance between each pair of points (an isometry). The corresponding group is called isometry group of X.

3. If instead angles are preserved, one speaks of conformal maps. Conformal maps give rise to Kleinian groups, for example.

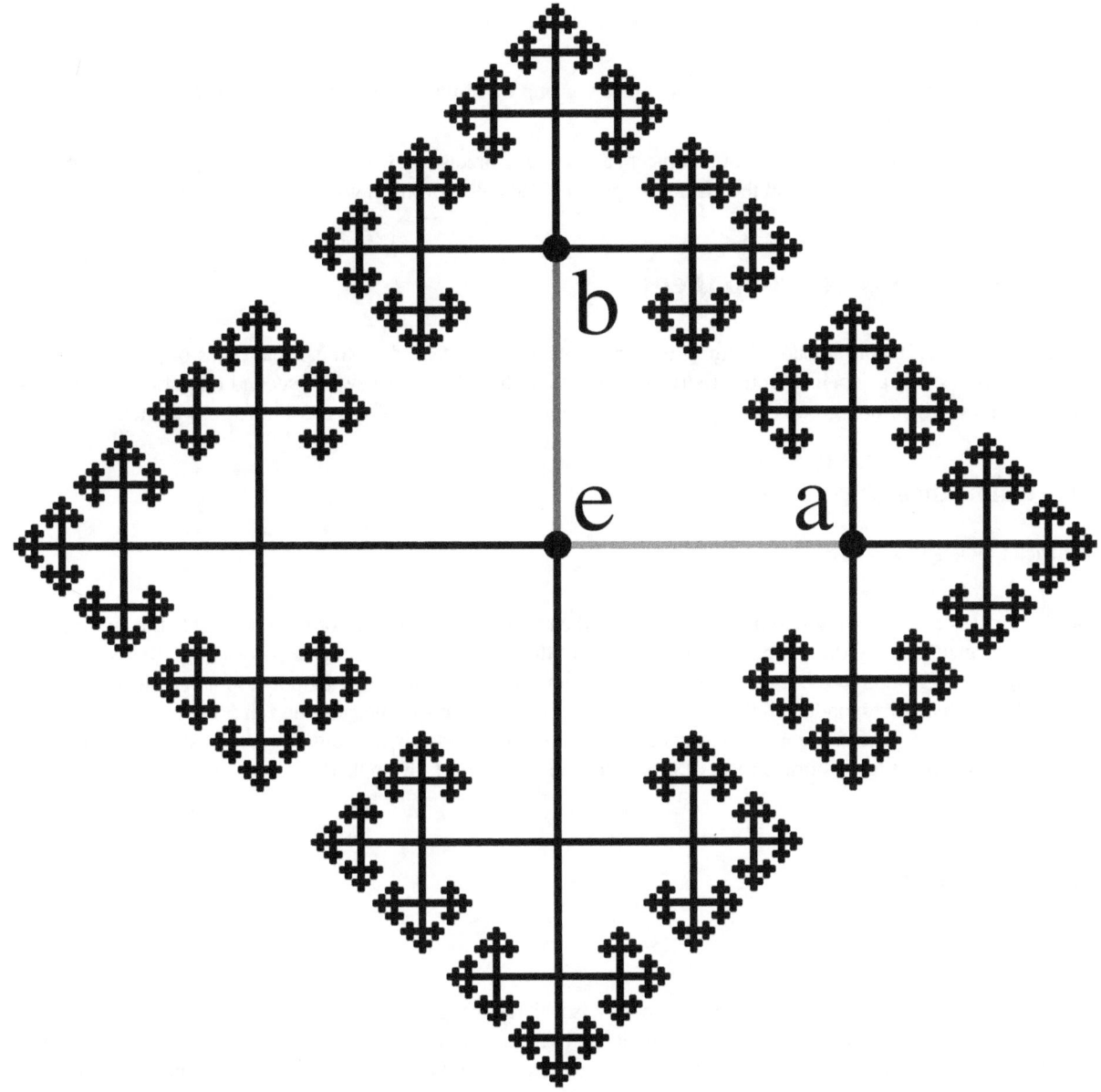

The Cayley graph of ⟨x, y | ⟩, the free group of rank 2.

4. Symmetries are not restricted to geometrical objects, but include algebraic objects as well. For instance, the equation

$$x^2 - 3 = 0$$

has the two solutions $+\sqrt{3}$, and $-\sqrt{3}$. In this case, the group that exchanges the two roots is the Galois group belonging to the equation. Every polynomial equation in one variable has a Galois group, that is a certain permutation group on its roots.

The axioms of a group formalize the essential aspects of symmetry. Symmetries form a group: they are closed because if you take a symmetry of an object, and then apply another symmetry, the result will still be a symmetry. The identity keeping the object fixed is always a symmetry of an object. Existence of inverses is guaranteed by undoing the symmetry and the associativity comes from the fact that symmetries are functions on a space, and composition of functions are associative.

Frucht's theorem says that every group is the symmetry group of some graph. So every abstract group is actually the symmetries of some explicit object.

The saying of "preserving the structure" of an object can be made precise by working in a category. Maps preserving the structure are then the morphisms, and the symmetry group is the automorphism group of the object in question.

2.4   Applications of group theory

Applications of group theory abound. Almost all structures in abstract algebra are special cases of groups. Rings, for example, can be viewed as abelian groups (corresponding to addition) together with a second operation (corresponding to multiplication). Therefore, group theoretic arguments underlie large parts of the theory of those entities.

2.4.1   Galois theory

Main article: Galois theory

Galois theory uses groups to describe the symmetries of the roots of a polynomial (or more precisely the automorphisms of the algebras generated by these roots). The fundamental theorem of Galois theory provides a link between algebraic field extensions and group theory. It gives an effective criterion for the solvability of polynomial equations in terms of the solvability of the corresponding Galois group. For example, S_5, the symmetric group in 5 elements, is not solvable which implies that the general quintic equation cannot be solved by radicals in the way equations of lower degree can. The theory, being one of the historical roots of group theory, is still fruitfully applied to yield new results in areas such as class field theory.

2.4.2   Algebraic topology

Main article: Algebraic topology

Algebraic topology is another domain which prominently associates groups to the objects the theory is interested in. There, groups are used to describe certain invariants of topological spaces. They are called "invariants" because they are defined in such a way that they do not change if the space is subjected to some deformation. For example, the fundamental group "counts" how many paths in the space are essentially different. The Poincaré conjecture, proved in 2002/2003 by Grigori Perelman, is a prominent application of this idea. The influence is not unidirectional, though. For example, algebraic topology makes use of Eilenberg–MacLane spaces which are spaces with prescribed homotopy groups. Similarly algebraic K-theory relies in a way on classifying spaces of groups. Finally, the name of the torsion subgroup of an infinite group shows the legacy of topology in group theory.

2.4.3   Algebraic geometry and cryptography

Main articles: Algebraic geometry and Cryptography

Algebraic geometry and cryptography likewise uses group theory in many ways. Abelian varieties have been introduced above. The presence of the group operation yields additional information which makes these varieties particularly accessible. They also often serve as a test for new conjectures.[9] The one-dimensional case, namely elliptic curves is studied in particular detail. They are both theoretically and practically intriguing.[10] Very large groups of prime order constructed in Elliptic-Curve Cryptography serve for public key cryptography. Cryptographical methods of this kind benefit from the flexibility of the geometric objects, hence their group structures, together with the complicated structure of these

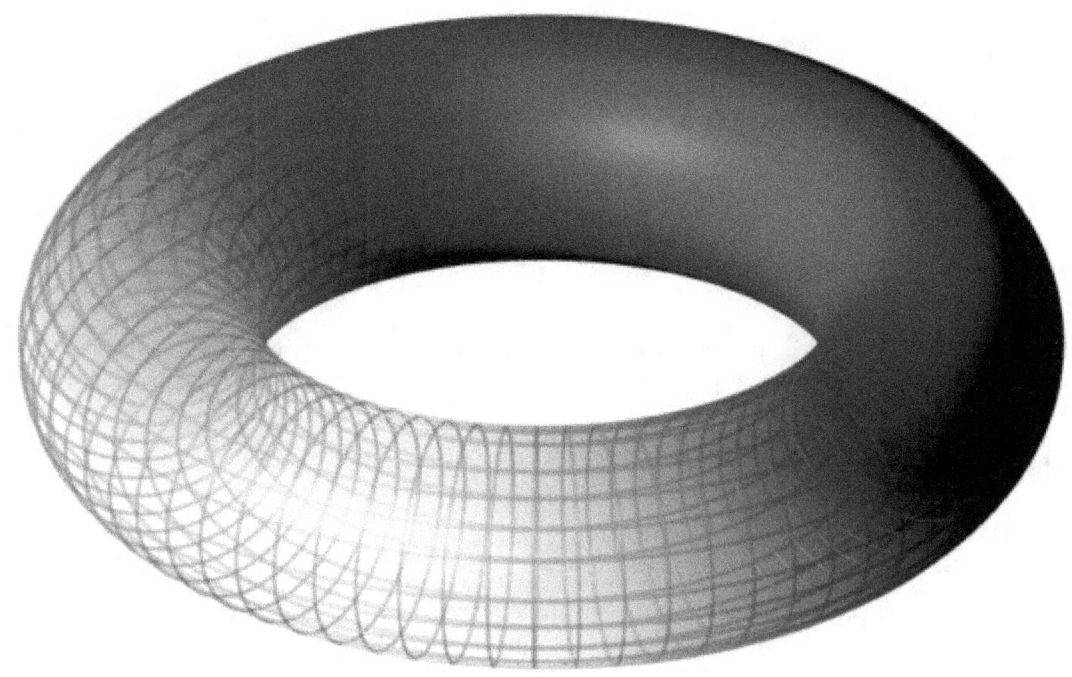

A torus. Its abelian group structure is induced from the map $C \to C/Z + \tau Z$, where τ is a parameter living in the upper half plane.

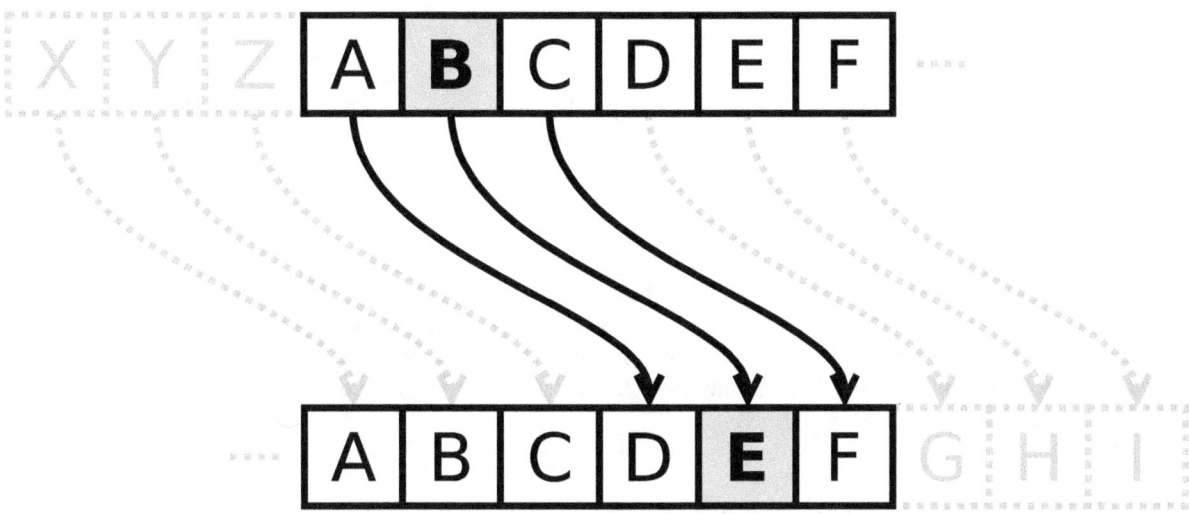

The cyclic group Z_{26} underlies Caesar's cipher.

groups, which make the discrete logarithm very hard to calculate. One of the earliest encryption protocols, Caesar's cipher, may also be interpreted as a (very easy) group operation. In another direction, toric varieties are algebraic varieties acted on by a torus. Toroidal embeddings have recently led to advances in algebraic geometry, in particular resolution of singularities.[11]

2.4.4   Algebraic number theory

Main article: Algebraic number theory

Algebraic number theory is a special case of group theory, thereby following the rules of the latter. For example, Euler's product formula

$$\sum_{n \geq 1} \frac{1}{n^s} = \prod_{p \text{ prime}} \frac{1}{1 - p^{-s}}$$

captures the fact that any integer decomposes in a unique way into primes. The failure of this statement for more general rings gives rise to class groups and regular primes, which feature in Kummer's treatment of Fermat's Last Theorem.

2.4.5   Harmonic analysis

Main article: Harmonic analysis

Analysis on Lie groups and certain other groups is called harmonic analysis. Haar measures, that is, integrals invariant under the translation in a Lie group, are used for pattern recognition and other image processing techniques.[12]

2.4.6   Combinatorics

In combinatorics, the notion of permutation group and the concept of group action are often used to simplify the counting of a set of objects; see in particular Burnside's lemma.

2.4.7   Music

The presence of the 12-periodicity in the circle of fifths yields applications of elementary group theory in musical set theory.

2.4.8   Physics

In physics, groups are important because they describe the symmetries which the laws of physics seem to obey. According to Noether's theorem, every continuous symmetry of a physical system corresponds to a conservation law of the system. Physicists are very interested in group representations, especially of Lie groups, since these representations often point the way to the "possible" physical theories. Examples of the use of groups in physics include the Standard Model, gauge theory, the Lorentz group, and the Poincaré group.

2.4.9   Chemistry and materials science

In chemistry and materials science, groups are used to classify crystal structures, regular polyhedra, and the symmetries of molecules. The assigned point groups can then be used to determine physical properties (such as polarity and chirality), spectroscopic properties (particularly useful for Raman spectroscopy and infrared spectroscopy), and to construct molecular orbitals.

Molecular symmetry is responsible for many physical and spectroscopic properties of compounds and provides relevant information about how chemical reactions occur. In order to assign a point group for any given molecule, it is necessary to find the set of symmetry operations present on it. The symmetry operation is an action, such as a rotation around an axis or a reflection through a mirror plane. In other words, it is an operation that moves the molecule such that it is

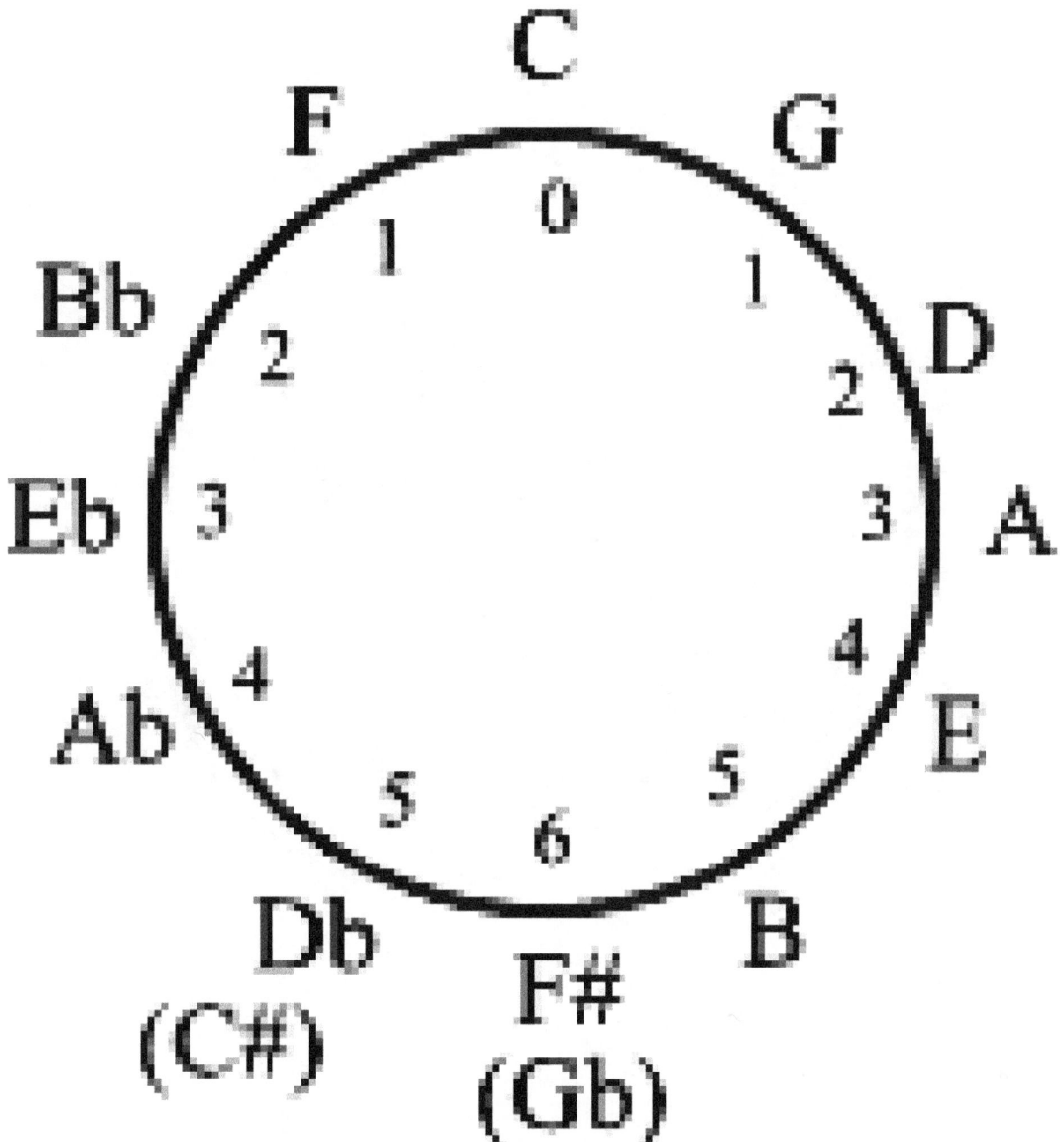

The circle of fifths may be endowed with a cyclic group structure

indistinguishable from the original configuration. In group theory, the rotation axes and mirror planes are called "symmetry elements". These elements can be a point, line or plane with respect to which the symmetry operation is carried out. The symmetry operations of a molecule determine the specific point group for this molecule.

In chemistry, there are five important symmetry operations. The identity operation (E) consists of leaving the molecule as it is. This is equivalent to any number of full rotations around any axis. This is a symmetry of all molecules, whereas the symmetry group of a chiral molecule consists of only the identity operation. Rotation around an axis (Cn) consists of rotating the molecule around a specific axis by a specific angle. For example, if a water molecule rotates $180°$ around the axis that passes through the oxygen atom and between the hydrogen atoms, it is in the same configuration as it started. In this case, $n = 2$, since applying it twice produces the identity operation. Other symmetry operations are: reflection,

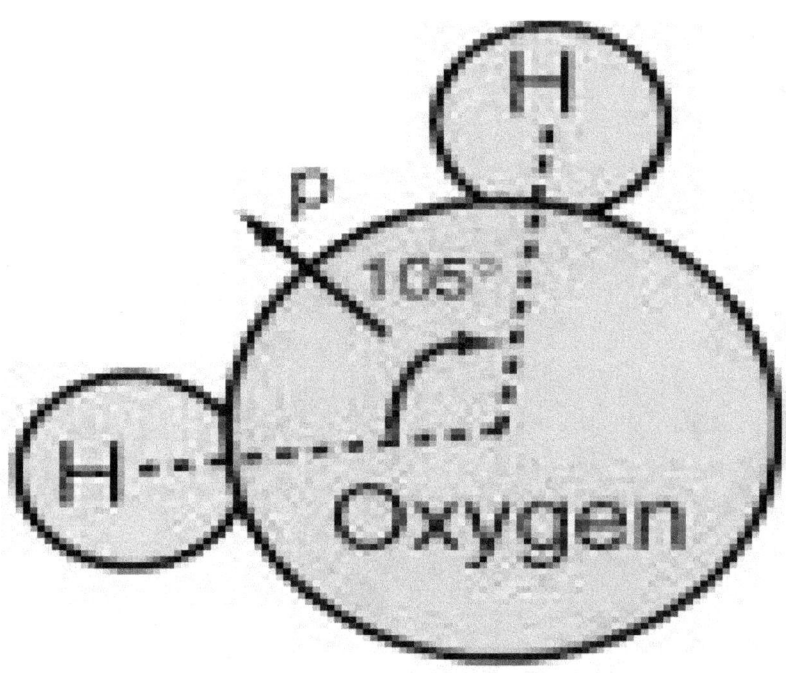

Water molecule with symmetry axis

inversion and improper rotation (rotation followed by reflection).[13]

2.4.10   Statistical Mechanics

Group theory can be used to resolve the incompleteness of the statistical interpretations of mechanics developed by Willard Gibbs, relating to the summing of an infinite number of probabilities to yield a meaningful solution[14]

2.5   History

Main article: History of group theory

Group theory has three main historical sources: number theory, the theory of algebraic equations, and geometry. The number-theoretic strand was begun by Leonhard Euler, and developed by Gauss's work on modular arithmetic and additive and multiplicative groups related to quadratic fields. Early results about permutation groups were obtained by Lagrange, Ruffini, and Abel in their quest for general solutions of polynomial equations of high degree. Évariste Galois coined the term "group" and established a connection, now known as Galois theory, between the nascent theory of groups and field theory. In geometry, groups first became important in projective geometry and, later, non-Euclidean geometry. Felix Klein's Erlangen program proclaimed group theory to be the organizing principle of geometry.

Galois, in the 1830s, was the first to employ groups to determine the solvability of polynomial equations. Arthur Cayley and Augustin Louis Cauchy pushed these investigations further by creating the theory of permutation groups. The second historical source for groups stems from geometrical situations. In an attempt to come to grips with possible geometries (such as euclidean, hyperbolic or projective geometry) using group theory, Felix Klein initiated the Erlangen programme. Sophus Lie, in 1884, started using groups (now called Lie groups) attached to analytic problems. Thirdly, groups were, at first implicitly and later explicitly, used in algebraic number theory.

The different scope of these early sources resulted in different notions of groups. The theory of groups was unified starting around 1880. Since then, the impact of group theory has been ever growing, giving rise to the birth of abstract algebra in the early 20th century, representation theory, and many more influential spin-off domains. The classification of finite simple groups is a vast body of work from the mid 20th century, classifying all the finite simple groups.

2.6 See also

- Group (mathematics)

- Glossary of group theory

- List of group theory topics

2.7 Notes

[1]   • Elwes, Richard, "An enormous theorem: the classification of finite simple groups," *Plus Magazine*, Issue 41, December 2006.

[2] This process of imposing extra structure has been formalized through the notion of a group object in a suitable category. Thus Lie groups are group objects in the category of differentiable manifolds and affine algebraic groups are group objects in the category of affine algebraic varieties.

[3] Such as group cohomology or equivariant K-theory.

[4] In particular, if the representation is faithful.

[5] Arthur Tresse (1893). "Sur les invariants différentiels des groupes continus de transformations". *Acta Mathematica* **18**: 1–88. doi:10.1007/bf02418270.

[6] Schupp & Lyndon 2001

[7] Writing $z = xy$, one has $G = ⟨z, y \mid z^3 = y⟩ = ⟨z⟩$.

[8] La Harpe 2000

[9] For example the Hodge conjecture (in certain cases).

[10] See the Birch-Swinnerton-Dyer conjecture, one of the millennium problems

[11] Abramovich, Dan; Karu, Kalle; Matsuki, Kenji; Wlodarczyk, Jaroslaw (2002), "Torification and factorization of birational maps", *Journal of the American Mathematical Society* **15** (3): 531–572, doi:10.1090/S0894-0347-02-00396-X, MR 1896232

[12] Lenz, Reiner (1990), *Group theoretical methods in image processing*, Lecture Notes in Computer Science **413**, Berlin, New York: Springer-Verlag, doi:10.1007/3-540-52290-5, ISBN 978-0-387-52290-6

[13] Shriver, D.F.; Atkins, P.W. Química Inorgânica, 3ª ed., Porto Alegre, Bookman, 2003.

[14] Norber Weiner, Cybernetics: Or Control and Communication in the Animal and the Machine, Ch 2

2.8   References

- Borel, Armand (1991), *Linear algebraic groups*, Graduate Texts in Mathematics **126** (2nd ed.), Berlin, New York: Springer-Verlag, ISBN 978-0-387-97370-8, MR 1102012

- Carter, Nathan C. (2009), *Visual group theory*, Classroom Resource Materials Series, Mathematical Association of America, ISBN 978-0-88385-757-1, MR 2504193

- Cannon, John J. (1969), "Computers in group theory: A survey", *Communications of the Association for Computing Machinery* **12**: 3–12, doi:10.1145/362835.362837, MR 0290613

- Frucht, R. (1939), "Herstellung von Graphen mit vorgegebener abstrakter Gruppe", *Compositio Mathematica* **6**: 239–50, ISSN 0010-437X

- Golubitsky, Martin; Stewart, Ian (2006), "Nonlinear dynamics of networks: the groupoid formalism", *Bull. Amer. Math. Soc. (N.S.)* **43** (03): 305–364, doi:10.1090/S0273-0979-06-01108-6, MR 2223010 Shows the advantage of generalising from group to groupoid.

- Judson, Thomas W. (1997), *Abstract Algebra: Theory and Applications* An introductory undergraduate text in the spirit of texts by Gallian or Herstein, covering groups, rings, integral domains, fields and Galois theory. Free downloadable PDF with open-source GFDL license.

- Kleiner, Israel (1986), "The evolution of group theory: a brief survey", *Mathematics Magazine* **59** (4): 195–215, doi:10.2307/2690312, ISSN 0025-570X, JSTOR 2690312, MR 863090

- La Harpe, Pierre de (2000), *Topics in geometric group theory*, University of Chicago Press, ISBN 978-0-226-31721-2

- Livio, M. (2005), *The Equation That Couldn't Be Solved: How Mathematical Genius Discovered the Language of Symmetry*, Simon & Schuster, ISBN 0-7432-5820-7 Conveys the practical value of group theory by explaining how it points to symmetries in physics and other sciences.

- Mumford, David (1970), *Abelian varieties*, Oxford University Press, ISBN 978-0-19-560528-0, OCLC 138290

- Ronan M., 2006. *Symmetry and the Monster*. Oxford University Press. ISBN 0-19-280722-6. For lay readers. Describes the quest to find the basic building blocks for finite groups.

- Rotman, Joseph (1994), *An introduction to the theory of groups*, New York: Springer-Verlag, ISBN 0-387-94285-8 A standard contemporary reference.

- Schupp, Paul E.; Lyndon, Roger C. (2001), *Combinatorial group theory*, Berlin, New York: Springer-Verlag, ISBN 978-3-540-41158-1

- Scott, W. R. (1987) [1964], *Group Theory*, New York: Dover, ISBN 0-486-65377-3 Inexpensive and fairly readable, but somewhat dated in emphasis, style, and notation.

- Shatz, Stephen S. (1972), *Profinite groups, arithmetic, and geometry*, Princeton University Press, ISBN 978-0-691-08017-8, MR 0347778

- Weibel, Charles A. (1994), *An introduction to homological algebra*, Cambridge Studies in Advanced Mathematics **38**, Cambridge University Press, ISBN 978-0-521-55987-4, OCLC 36131259, MR 1269324

2.9 External links

- History of the abstract group concept

- Higher dimensional group theory This presents a view of group theory as level one of a theory which extends in all dimensions, and has applications in homotopy theory and to higher dimensional nonabelian methods for local-to-global problems.

- Plus teacher and student package: Group Theory This package brings together all the articles on group theory from *Plus*, the online mathematics magazine produced by the Millennium Mathematics Project at the University of Cambridge, exploring applications and recent breakthroughs, and giving explicit definitions and examples of groups.

- US Naval Academy group theory guide A general introduction to group theory with exercises written by Tony Gaglione.

Chapter 3

Lie group

In mathematics, a **Lie group** /'liː/ is a group that is also a differentiable manifold, with the property that the group operations are compatible with the smooth structure. Lie groups are named after Sophus Lie, who laid the foundations of the theory of continuous transformation groups. The term *groupes de Lie* first appeared in French in 1893 in the thesis of Lie's student Arthur Tresse, page 3.[1]

Lie groups represent the best-developed theory of continuous symmetry of mathematical objects and structures, which makes them indispensable tools for many parts of contemporary mathematics, as well as for modern theoretical physics. They provide a natural framework for analysing the continuous symmetries of differential equations (differential Galois theory), in much the same way as permutation groups are used in Galois theory for analysing the discrete symmetries of algebraic equations. An extension of Galois theory to the case of continuous symmetry groups was one of Lie's principal motivations.

3.1 Overview

Lie groups are smooth[Note 1] differentiable manifolds and as such can be studied using differential calculus, in contrast with the case of more general topological groups. One of the key ideas in the theory of Lie groups is to replace the *global* object, the group, with its *local* or linearized version, which Lie himself called its "infinitesimal group" and which has since become known as its Lie algebra.

Lie groups play an enormous role in modern geometry, on several different levels. Felix Klein argued in his Erlangen program that one can consider various "geometries" by specifying an appropriate transformation group that leaves certain geometric properties invariant. Thus Euclidean geometry corresponds to the choice of the group E(3) of distance-preserving transformations of the Euclidean space $\mathbf{R}^3$, conformal geometry corresponds to enlarging the group to the conformal group, whereas in projective geometry one is interested in the properties invariant under the projective group. This idea later led to the notion of a G-structure, where G is a Lie group of "local" symmetries of a manifold. On a "global" level, whenever a Lie group acts on a geometric object, such as a Riemannian or a symplectic manifold, this action provides a measure of rigidity and yields a rich algebraic structure. The presence of continuous symmetries expressed via a Lie group action on a manifold places strong constraints on its geometry and facilitates analysis on the manifold. Linear actions of Lie groups are especially important, and are studied in representation theory.

In the 1940s–1950s, Ellis Kolchin, Armand Borel, and Claude Chevalley realised that many foundational results concerning Lie groups can be developed completely algebraically, giving rise to the theory of algebraic groups defined over an arbitrary field. This insight opened new possibilities in pure algebra, by providing a uniform construction for most finite simple groups, as well as in algebraic geometry. The theory of automorphic forms, an important branch of modern number theory, deals extensively with analogues of Lie groups over adele rings; p-adic Lie groups play an important role, via their connections with Galois representations in number theory.

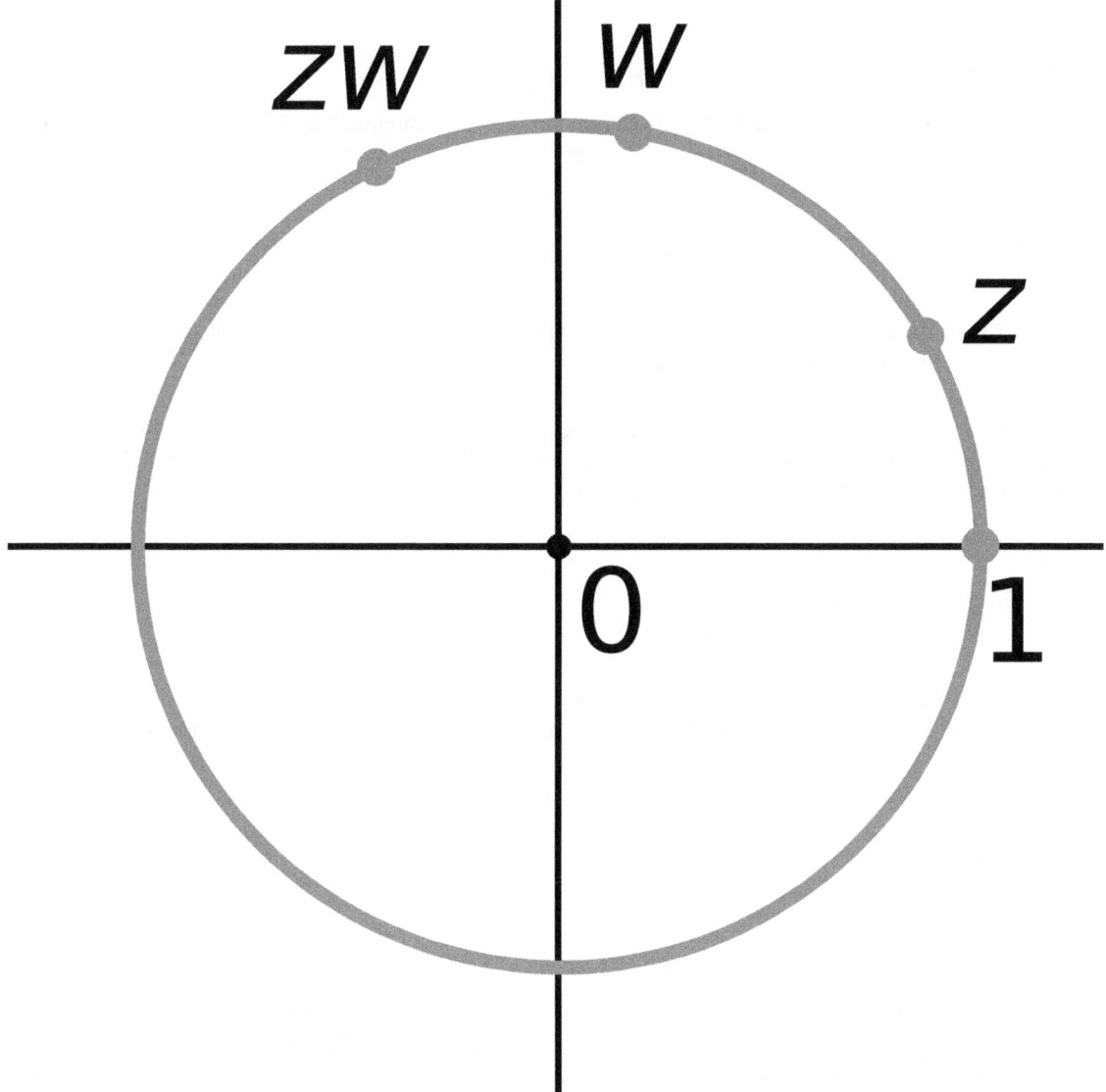

The circle of center 0 and radius 1 in the complex plane is a Lie group with complex multiplication.

3.2 Definitions and examples

A **real Lie group** is a group that is also a finite-dimensional real smooth manifold, in which the group operations of multiplication and inversion are smooth maps. Smoothness of the group multiplication

$$\mu : G \times G \to G \quad \mu(x, y) = xy$$

means that μ is a smooth mapping of the product manifold $G \times G$ into G. These two requirements can be combined to the single requirement that the mapping

$$(x, y) \mapsto x^{-1}y$$

be a smooth mapping of the product manifold into *G*.

3.2.1 First examples

- The 2×2 real invertible matrices form a group under multiplication, denoted by GL(2, **R**) or by GL2(**R**):

$$\mathrm{GL}(2, \mathbf{R}) = \left\{ A = \begin{pmatrix} a & b \\ c & d \end{pmatrix} : \det A = ad - bc \neq 0 \right\}.$$

  This is a four-dimensional noncompact real Lie group. This group is disconnected; it has two connected components corresponding to the positive and negative values of the determinant.

- The rotation matrices form a subgroup of GL(2, **R**), denoted by SO(2, **R**). It is a Lie group in its own right: specifically, a one-dimensional compact connected Lie group which is diffeomorphic to the circle. Using the rotation angle φ as a parameter, this group can be parametrized as follows:

$$\mathrm{SO}(2, \mathbf{R}) = \left\{ \begin{pmatrix} \cos\varphi & -\sin\varphi \\ \sin\varphi & \cos\varphi \end{pmatrix} : \varphi \in \mathbf{R}/2\pi\mathbf{Z} \right\}.$$

  Addition of the angles corresponds to multiplication of the elements of SO(2, **R**), and taking the opposite angle corresponds to inversion. Thus both multiplication and inversion are differentiable maps.

- The orthogonal group also forms an interesting example of a Lie group.

All of the previous examples of Lie groups fall within the class of classical groups.

3.2.2 Related concepts

A **complex Lie group** is defined in the same way using complex manifolds rather than real ones (example: SL(2, **C**)), and similarly, using an alternate metric completion of **Q**, one can define a *p*-**adic Lie group** over the *p*-adic numbers, a topological group in which each point has a *p*-adic neighborhood. Hilbert's fifth problem asked whether replacing differentiable manifolds with topological or analytic ones can yield new examples. The answer to this question turned out to be negative: in 1952, Gleason, Montgomery and Zippin showed that if *G* is a topological manifold with continuous group operations, then there exists exactly one analytic structure on *G* which turns it into a Lie group (see also Hilbert–Smith conjecture). If the underlying manifold is allowed to be infinite-dimensional (for example, a Hilbert manifold), then one arrives at the notion of an infinite-dimensional Lie group. It is possible to define analogues of many Lie groups over finite fields, and these give most of the examples of finite simple groups.

The language of category theory provides a concise definition for Lie groups: a Lie group is a group object in the category of smooth manifolds. This is important, because it allows generalization of the notion of a Lie group to Lie supergroups.

3.3 More examples of Lie groups

See also: Table of Lie groups and List of simple Lie groups

Lie groups occur in abundance throughout mathematics and physics. Matrix groups or algebraic groups are (roughly) groups of matrices (for example, orthogonal and symplectic groups), and these give most of the more common examples of Lie groups.

3.3.1 Examples with a specific number of dimensions

- The circle group S^1 consisting of angles mod 2π under addition or, alternatively, the complex numbers with absolute value 1 under multiplication. This is a one-dimensional compact connected abelian Lie group.

- The 3-sphere S^3 forms a Lie group by identification with the set of quaternions of unit norm, called versors. The only other spheres that admit the structure of a Lie group are the 0-sphere S^0 (real numbers with absolute value 1) and the circle S^1 (complex numbers with absolute value 1). For example, for even $n > 1$, S^n is not a Lie group because it does not admit a nonvanishing vector field and so *a fortiori* cannot be parallelizable as a differentiable manifold. Of the spheres only S^0, S^1, S^3, and S^7 are parallelizable. The last carries the structure of a Lie quasigroup (a nonassociative group), which can be identified with the set of unit octonions.

- The (3-dimensional) metaplectic group is a double cover of SL(2, $\mathbf{R}$) playing an important role in the theory of modular forms. It is a connected Lie group that cannot be faithfully represented by matrices of finite size, i.e., a nonlinear group.

- The Heisenberg group is a connected nilpotent Lie group of dimension 3, playing a key role in quantum mechanics.

- The Lorentz group is a 6-dimensional Lie group of linear isometries of the Minkowski space.

- The Poincaré group is a 10-dimensional Lie group of affine isometries of the Minkowski space.

- The group U(1)×SU(2)×SU(3) is a Lie group of dimension 1+3+8=12 that is the gauge group of the Standard Model in particle physics. The dimensions of the factors correspond to the 1 photon + 3 vector bosons + 8 gluons of the standard model

- The exceptional Lie groups of types G_2, F_4, E_6, E_7, E_8 have dimensions 14, 52, 78, 133, and 248. Along with the A-B-C-D series of simple Lie groups, the exceptional groups complete the list of simple Lie groups. There is also a Lie group named $E_7\frac{1}{2}$ of dimension 190, but it is not a *simple* Lie group.

3.3.2 Examples with n dimensions

- Euclidean space $\mathbf{R}^n$ with ordinary vector addition as the group operation becomes an n-dimensional noncompact abelian Lie group.

- The Euclidean group E(n, $\mathbf{R}$) is the Lie group of all Euclidean motions, i.e., isometric affine maps, of n-dimensional Euclidean space $\mathbf{R}^n$.

- The orthogonal group O(n, $\mathbf{R}$), consisting of all $n \times n$ orthogonal matrices with real entries is an $n(n-1)/2$-dimensional Lie group. This group is disconnected, but it has a connected subgroup SO(n, $\mathbf{R}$) of the same dimension consisting of orthogonal matrices of determinant 1, called the special orthogonal group (for $n = 3$, the rotation group SO(3)).

- The unitary group U(n) consisting of $n \times n$ unitary matrices (with complex entries) is a compact connected Lie group of dimension n^2. Unitary matrices of determinant 1 form a closed connected subgroup of dimension $n^2 - 1$ denoted SU(n), the special unitary group.

- Spin groups are double covers of the special orthogonal groups, used for studying fermions in quantum field theory (among other things).

- The group GL(n, $\mathbf{R}$) of invertible matrices (under matrix multiplication) is a Lie group of dimension n^2, called the general linear group. It has a closed connected subgroup SL(n, $\mathbf{R}$), the special linear group, consisting of matrices of determinant 1 which is also a Lie group.

- The symplectic group Sp($2n$, $\mathbf{R}$) consists of all $2n \times 2n$ matrices preserving a *symplectic form* on $\mathbf{R}^{2n}$. It is a connected Lie group of dimension $2n^2 + n$.

- The group of invertible upper triangular n by n matrices is a solvable Lie group of dimension $n(n + 1)/2$. (cf. Borel subgroup)

- The A-series, B-series, C-series and D-series, whose elements are denoted by An, Bn, Cn, and Dn, are infinite families of simple Lie groups.

3.3.3 Constructions

There are several standard ways to form new Lie groups from old ones:

- The product of two Lie groups is a Lie group.

- Any topologically closed subgroup of a Lie group is a Lie group. This is known as the Closed subgroup theorem or **Cartan's theorem**.

- The quotient of a Lie group by a closed normal subgroup is a Lie group.

- The universal cover of a connected Lie group is a Lie group. For example, the group **R** is the universal cover of the circle group S^1. In fact any covering of a differentiable manifold is also a differentiable manifold, but by specifying *universal* cover, one guarantees a group structure (compatible with its other structures).

3.3.4 Related notions

Some examples of groups that are *not* Lie groups (except in the trivial sense that any group can be viewed as a 0-dimensional Lie group, with the discrete topology), are:

- Infinite-dimensional groups, such as the additive group of an infinite-dimensional real vector space. These are not Lie groups as they are not *finite-dimensional* manifolds.

- Some totally disconnected groups, such as the Galois group of an infinite extension of fields, or the additive group of the *p*-adic numbers. These are not Lie groups because their underlying spaces are not real manifolds. (Some of these groups are "*p*-adic Lie groups".) In general, only topological groups having similar local properties to $\mathbf{R}^n$ for some positive integer n can be Lie groups (of course they must also have a differentiable structure).

3.4 Basic concepts

3.4.1 The Lie algebra associated with a Lie group

Main article: Lie group–Lie algebra correspondence

To every Lie group we can associate a Lie algebra whose underlying vector space is the tangent space of the Lie group at the identity element and which completely captures the local structure of the group. Informally we can think of elements of the Lie algebra as elements of the group that are "infinitesimally close" to the identity, and the Lie bracket of the Lie algebra is related to the commutator of two such infinitesimal elements. Before giving the abstract definition we give a few examples:

- The Lie algebra of the vector space $\mathbf{R}^n$ is just $\mathbf{R}^n$ with the Lie bracket given by
$[A, B] = 0$.
(In general the Lie bracket of a connected Lie group is always 0 if and only if the Lie group is abelian.)

- The Lie algebra of the general linear group GL(n, **R**) of invertible matrices is the vector space M(n, **R**) of square matrices with the Lie bracket given by
$[A, B] = AB - BA$.
If G is a closed subgroup of GL(n, **R**) then the Lie algebra of G can be thought of informally as the matrices m

of M(n, **R**) such that $1 + \varepsilon m$ is in G, where ε is an infinitesimal positive number with $\varepsilon^2 = 0$ (of course, no such real number ε exists). For example, the orthogonal group O(n, **R**) consists of matrices A with $AA^T = 1$, so the Lie algebra consists of the matrices m with $(1 + \varepsilon m)(1 + \varepsilon m)^T = 1$, which is equivalent to $m + m^T = 0$ because $\varepsilon^2 = 0$.

- Formally, when working over the reals, as here, this is accomplished by considering the limit as $\varepsilon \to 0$; but the "infinitesimal" language generalizes directly to Lie groups over general rings.

The concrete definition given above is easy to work with, but has some minor problems: to use it we first need to represent a Lie group as a group of matrices, but not all Lie groups can be represented in this way, and it is not obvious that the Lie algebra is independent of the representation we use. To get around these problems we give the general definition of the Lie algebra of a Lie group (in 4 steps):

1. Vector fields on any smooth manifold M can be thought of as derivations X of the ring of smooth functions on the manifold, and therefore form a Lie algebra under the Lie bracket $[X, Y] = XY - YX$, because the Lie bracket of any two derivations is a derivation.

2. If G is any group acting smoothly on the manifold M, then it acts on the vector fields, and the vector space of vector fields fixed by the group is closed under the Lie bracket and therefore also forms a Lie algebra.

3. We apply this construction to the case when the manifold M is the underlying space of a Lie group G, with G acting on $G = M$ by left translations $Lg(h) = gh$. This shows that the space of left invariant vector fields (vector fields satisfying $Lg*Xh = Xgh$ for every h in G, where $Lg*$ denotes the differential of Lg) on a Lie group is a Lie algebra under the Lie bracket of vector fields.

4. Any tangent vector at the identity of a Lie group can be extended to a left invariant vector field by left translating the tangent vector to other points of the manifold. Specifically, the left invariant extension of an element v of the tangent space at the identity is the vector field defined by $v^\wedge g = Lg*v$. This identifies the tangent space TeG at the identity with the space of left invariant vector fields, and therefore makes the tangent space at the identity into a Lie algebra, called the Lie algebra of G, usually denoted by a Fraktur $\mathfrak{g}$. Thus the Lie bracket on $\mathfrak{g}$ is given explicitly by $[v, w] = [v^\wedge, w^\wedge]e$.

This Lie algebra $\mathfrak{g}$ is finite-dimensional and it has the same dimension as the manifold G. The Lie algebra of G determines G up to "local isomorphism", where two Lie groups are called **locally isomorphic** if they look the same near the identity element. Problems about Lie groups are often solved by first solving the corresponding problem for the Lie algebras, and the result for groups then usually follows easily. For example, simple Lie groups are usually classified by first classifying the corresponding Lie algebras.

We could also define a Lie algebra structure on Te using right invariant vector fields instead of left invariant vector fields. This leads to the same Lie algebra, because the inverse map on G can be used to identify left invariant vector fields with right invariant vector fields, and acts as -1 on the tangent space Te.

The Lie algebra structure on Te can also be described as follows: the commutator operation

$$(x, y) \to xyx^{-1}y^{-1}$$

on $G \times G$ sends (e, e) to e, so its derivative yields a bilinear operation on TeG. This bilinear operation is actually the zero map, but the second derivative, under the proper identification of tangent spaces, yields an operation that satisfies the axioms of a Lie bracket, and it is equal to twice the one defined through left-invariant vector fields.

3.4.2 Homomorphisms and isomorphisms

If G and H are Lie groups, then a Lie group homomorphism $f : G \to H$ is a smooth group homomorphism. In the case of complex Lie groups, such a homomorphism is required to be a holomorphic map. However, these requirements are a bit stringent; over real or complex numbers, every continuous homomorphism between Lie groups turns out to be (real or complex) analytic.

The composition of two Lie homomorphisms is again a homomorphism, and the class of all Lie groups, together with these morphisms, forms a category. Moreover, every Lie group homomorphism induces a homomorphism between the corresponding Lie algebras. Let $\phi\colon G \to H$ be a Lie group homomorphism and let ϕ_* be its derivative at the identity. If we identify the Lie algebras of G and H with their tangent spaces at the identity elements then ϕ_* is a map between the corresponding Lie algebras:

$$\phi_* \colon \mathfrak{g} \to \mathfrak{h}$$

One can show that ϕ_* is actually a Lie algebra homomorphism (meaning that it is a linear map which preserves the Lie bracket). In the language of category theory, we then have a covariant functor from the category of Lie groups to the category of Lie algebras which sends a Lie group to its Lie algebra and a Lie group homomorphism to its derivative at the identity.

Two Lie groups are called *isomorphic* if there exists a bijective homomorphism between them whose inverse is also a Lie group homomorphism. Equivalently, it is a diffeomorphism which is also a group homomorphism.

Ado's theorem says every finite-dimensional Lie algebra is isomorphic to a matrix Lie algebra. For every finite-dimensional matrix Lie algebra, there is a linear group (matrix Lie group) with this algebra as its Lie algebra. So every abstract Lie algebra is the Lie algebra of some (linear) Lie group.

The *global structure* of a Lie group is not determined by its Lie algebra; for example, if Z is any discrete subgroup of the center of G then G and G/Z have the same Lie algebra (see the table of Lie groups for examples). A *connected* Lie group is simple, semisimple, solvable, nilpotent, or abelian if and only if its Lie algebra has the corresponding property.

If we require that the Lie group be simply connected, then the global structure is determined by its Lie algebra: for every finite-dimensional Lie algebra $\mathfrak{g}$ over $\mathbf{F}$ there is a simply connected Lie group G with $\mathfrak{g}$ as Lie algebra, unique up to isomorphism. Moreover every homomorphism between Lie algebras lifts to a unique homomorphism between the corresponding simply connected Lie groups.

3.4.3   The exponential map

Main article: Exponential map (Lie theory)

The exponential map from the Lie algebra $\mathrm{M}(n, \mathbf{R})$ of the general linear group $\mathrm{GL}(n, \mathbf{R})$ to $\mathrm{GL}(n, \mathbf{R})$ is defined by the usual power series:

$$\exp(A) = 1 + A + \frac{A^2}{2!} + \frac{A^3}{3!} + \cdots$$

for matrices A. If G is any subgroup of $\mathrm{GL}(n, \mathbf{R})$, then the exponential map takes the Lie algebra of G into G, so we have an exponential map for all matrix groups.

The definition above is easy to use, but it is not defined for Lie groups that are not matrix groups, and it is not clear that the exponential map of a Lie group does not depend on its representation as a matrix group. We can solve both problems using a more abstract definition of the exponential map that works for all Lie groups, as follows.

Every vector v in $\mathfrak{g}$ determines a linear map from $\mathbf{R}$ to $\mathfrak{g}$ taking 1 to v, which can be thought of as a Lie algebra homomorphism. Because $\mathbf{R}$ is the Lie algebra of the simply connected Lie group $\mathbf{R}$, this induces a Lie group homomorphism $c\colon \mathbf{R} \to G$ so that

$$c(s + t) = c(s)c(t)$$

for all s and t. The operation on the right hand side is the group multiplication in G. The formal similarity of this formula with the one valid for the exponential function justifies the definition

$\exp(v) = c(1)$.

This is called the **exponential map**, and it maps the Lie algebra $\mathfrak{g}$ into the Lie group G. It provides a diffeomorphism between a neighborhood of 0 in $\mathfrak{g}$ and a neighborhood of e in G. This exponential map is a generalization of the exponential function for real numbers (because **R** is the Lie algebra of the Lie group of positive real numbers with multiplication), for complex numbers (because **C** is the Lie algebra of the Lie group of non-zero complex numbers with multiplication) and for matrices (because M(n, **R**) with the regular commutator is the Lie algebra of the Lie group GL(n, **R**) of all invertible matrices).

Because the exponential map is surjective on some neighbourhood N of e, it is common to call elements of the Lie algebra **infinitesimal generators** of the group G. The subgroup of G generated by N is the identity component of G.

The exponential map and the Lie algebra determine the *local group structure* of every connected Lie group, because of the Baker–Campbell–Hausdorff formula: there exists a neighborhood U of the zero element of $\mathfrak{g}$, such that for u, v in U we have

$$\exp(u)\,\exp(v) = \exp\left(u + v + \tfrac{1}{2}[u,v] + \tfrac{1}{12}[\,[u,v],v] - \tfrac{1}{12}[\,[u,v],u] - \cdots\right),$$

where the omitted terms are known and involve Lie brackets of four or more elements. In case u and v commute, this formula reduces to the familiar exponential law $\exp(u)\,\exp(v) = \exp(u+v)$.

The exponential map relates Lie group homomorphisms. That is, if $\phi : G \to H$ is a Lie group homomorphism and $\phi_* : \mathfrak{g} \to \mathfrak{h}$ the induced map on the corresponding Lie algebras, then for all $x \in \mathfrak{g}$ we have

$$\phi(\exp(x)) = \exp(\phi_*(x)).$$

In other words the following diagram commutes,[Note 2]

(In short, exp is a natural transformation from the functor Lie to the identity functor on the category of Lie groups.)

The exponential map from the Lie algebra to the Lie group is not always onto, even if the group is connected (though it does map onto the Lie group for connected groups that are either compact or nilpotent). For example, the exponential map of SL(2, **R**) is not surjective. Also, exponential map is not surjective nor injective for infinite-dimensional (see below) Lie groups modelled on C^∞ Fréchet space, even from arbitrary small neighborhood of 0 to corresponding neighborhood of 1.

See also: derivative of the exponential map and normal coordinates.

3.4.4 Lie subgroup

A **Lie subgroup** H of a Lie group G is a Lie group that is a subset of G and such that the inclusion map from H to G is an injective immersion and group homomorphism. According to Cartan's theorem, a closed subgroup of G admits a unique smooth structure which makes it an embedded Lie subgroup of G—i.e. a Lie subgroup such that the inclusion map is a smooth embedding.

Examples of non-closed subgroups are plentiful; for example take G to be a torus of dimension ≥ 2, and let H be a one-parameter subgroup of *irrational slope*, i.e. one that winds around in G. Then there is a Lie group homomorphism $\varphi : $ **R** $\to G$ with H as its image. The closure of H will be a sub-torus in G.

In terms of the exponential map of G, in general, only some of the Lie subalgebras of the Lie algebra g of G correspond to closed Lie subgroups H of G. There is no criterion solely based on the structure of g which determines which those are.

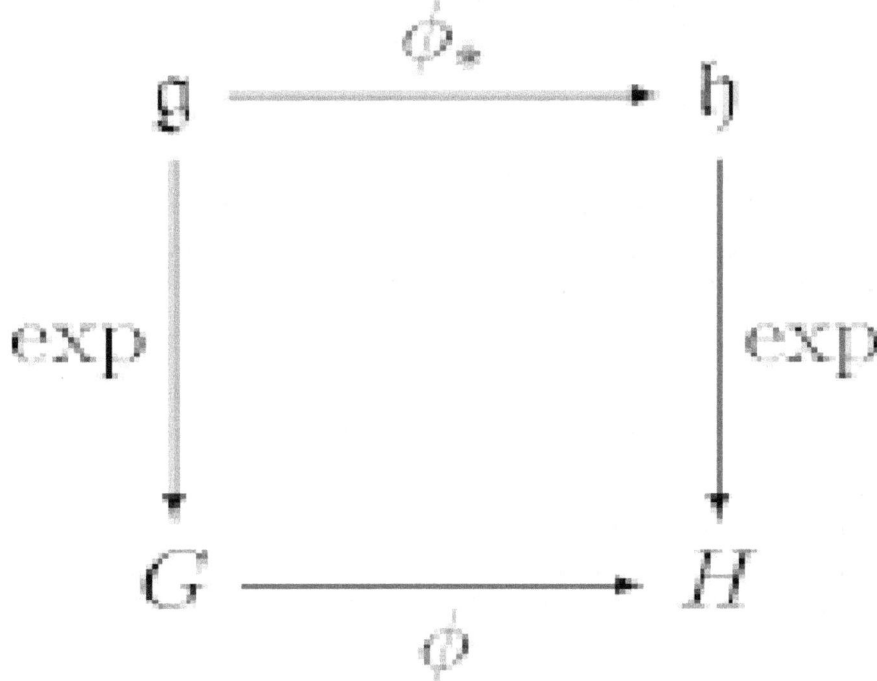

3.5   Early history

According to the most authoritative source on the early history of Lie groups (Hawkins, p. 1), Sophus Lie himself considered the winter of 1873–1874 as the birth date of his theory of continuous groups. Hawkins, however, suggests that it was "Lie's prodigious research activity during the four-year period from the fall of 1869 to the fall of 1873" that led to the theory's creation (*ibid*). Some of Lie's early ideas were developed in close collaboration with Felix Klein. Lie met with Klein every day from October 1869 through 1872: in Berlin from the end of October 1869 to the end of February 1870, and in Paris, Göttingen and Erlangen in the subsequent two years (*ibid*, p. 2). Lie stated that all of the principal results were obtained by 1884. But during the 1870s all his papers (except the very first note) were published in Norwegian journals, which impeded recognition of the work throughout the rest of Europe (*ibid*, p. 76). In 1884 a young German mathematician, Friedrich Engel, came to work with Lie on a systematic treatise to expose his theory of continuous groups. From this effort resulted the three-volume *Theorie der Transformationsgruppen*, published in 1888, 1890, and 1893.

Lie's ideas did not stand in isolation from the rest of mathematics. In fact, his interest in the geometry of differential equations was first motivated by the work of Carl Gustav Jacobi, on the theory of partial differential equations of first order and on the equations of classical mechanics. Much of Jacobi's work was published posthumously in the 1860s, generating enormous interest in France and Germany (Hawkins, p. 43). Lie's *idée fixe* was to develop a theory of symmetries of differential equations that would accomplish for them what Évariste Galois had done for algebraic equations: namely, to classify them in terms of group theory. Lie and other mathematicians showed that the most important equations for special functions and orthogonal polynomials tend to arise from group theoretical symmetries. In Lie's early work, the idea was to construct a theory of *continuous groups*, to complement the theory of discrete groups that had developed in the theory of modular forms, in the hands of Felix Klein and Henri Poincaré. The initial application that Lie had in mind was to the theory of differential equations. On the model of Galois theory and polynomial equations, the driving conception was of a theory capable of unifying, by the study of symmetry, the whole area of ordinary differential equations. However, the hope that Lie Theory would unify the entire field of ordinary differential equations was not fulfilled. Symmetry methods for ODEs continue to be studied, but do not dominate the subject. There is a differential Galois theory, but it was developed by others, such as Picard and Vessiot, and it provides a theory of quadratures, the indefinite integrals required to express solutions.

Additional impetus to consider continuous groups came from ideas of Bernhard Riemann, on the foundations of geometry, and their further development in the hands of Klein. Thus three major themes in 19th century mathematics were combined by Lie in creating his new theory: the idea of symmetry, as exemplified by Galois through the algebraic notion of a group; geometric theory and the explicit solutions of differential equations of mechanics, worked out by Poisson and Jacobi; and the new understanding of geometry that emerged in the works of Plücker, Möbius, Grassmann and others, and culminated in Riemann's revolutionary vision of the subject.

Although today Sophus Lie is rightfully recognized as the creator of the theory of continuous groups, a major stride in the development of their structure theory, which was to have a profound influence on subsequent development of mathematics, was made by Wilhelm Killing, who in 1888 published the first paper in a series entitled *Die Zusammensetzung der stetigen endlichen Transformationsgruppen* (*The composition of continuous finite transformation groups*) (Hawkins, p. 100). The work of Killing, later refined and generalized by Élie Cartan, led to classification of semisimple Lie algebras, Cartan's theory of symmetric spaces, and Hermann Weyl's description of representations of compact and semisimple Lie groups using highest weights.

In 1900 David Hilbert challenged Lie theorists with his Fifth Problem presented at the International Congress of Mathematicians in Paris.

Weyl brought the early period of the development of the theory of Lie groups to fruition, for not only did he classify irreducible representations of semisimple Lie groups and connect the theory of groups with quantum mechanics, but he also put Lie's theory itself on firmer footing by clearly enunciating the distinction between Lie's *infinitesimal groups* (i.e., Lie algebras) and the Lie groups proper, and began investigations of topology of Lie groups.[2] The theory of Lie groups was systematically reworked in modern mathematical language in a monograph by Claude Chevalley.

3.6 The concept of a Lie group, and possibilities of classification

Lie groups may be thought of as smoothly varying families of symmetries. Examples of symmetries include rotation about an axis. What must be understood is the nature of 'small' transformations, e.g., rotations through tiny angles, that link nearby transformations. The mathematical object capturing this structure is called a Lie algebra (Lie himself called them "infinitesimal groups"). It can be defined because Lie groups are manifolds, so have tangent spaces at each point.

The Lie algebra of any compact Lie group (very roughly: one for which the symmetries form a bounded set) can be decomposed as a direct sum of an abelian Lie algebra and some number of simple ones. The structure of an abelian Lie algebra is mathematically uninteresting (since the Lie bracket is identically zero); the interest is in the simple summands. Hence the question arises: what are the simple Lie algebras of compact groups? It turns out that they mostly fall into four infinite families, the "classical Lie algebras" A_n, B_n, C_n and D_n, which have simple descriptions in terms of symmetries of Euclidean space. But there are also just five "exceptional Lie algebras" that do not fall into any of these families. E_8 is the largest of these.

Lie groups are classified according to their algebraic properties (simple, semisimple, solvable, nilpotent, abelian), their connectedness (connected or simply connected) and their compactness.

- Compact Lie groups are all known: they are finite central quotients of a product of copies of the circle group S^1 and simple compact Lie groups (which correspond to connected Dynkin diagrams).

- Any simply connected solvable Lie group is isomorphic to a closed subgroup of the group of invertible upper triangular matrices of some rank, and any finite-dimensional irreducible representation of such a group is 1-dimensional. Solvable groups are too messy to classify except in a few small dimensions.

- Any simply connected nilpotent Lie group is isomorphic to a closed subgroup of the group of invertible upper triangular matrices with 1's on the diagonal of some rank, and any finite-dimensional irreducible representation of such a group is 1-dimensional. Like solvable groups, nilpotent groups are too messy to classify except in a few small dimensions.

- Simple Lie groups are sometimes defined to be those that are simple as abstract groups, and sometimes defined to be connected Lie groups with a simple Lie algebra. For example, SL(2, **R**) is simple according to the second definition but not according to the first. They have all been classified (for either definition).

- Semisimple Lie groups are Lie groups whose Lie algebra is a product of simple Lie algebras.[3] They are central extensions of products of simple Lie groups.

The identity component of any Lie group is an open normal subgroup, and the quotient group is a discrete group. The universal cover of any connected Lie group is a simply connected Lie group, and conversely any connected Lie group is a quotient of a simply connected Lie group by a discrete normal subgroup of the center. Any Lie group G can be decomposed into discrete, simple, and abelian groups in a canonical way as follows. Write

G_{con} for the connected component of the identity

G_{sol} for the largest connected normal solvable subgroup

G_{nil} for the largest connected normal nilpotent subgroup

so that we have a sequence of normal subgroups

$$1 \subseteq G_{\mathrm{nil}} \subseteq G_{\mathrm{sol}} \subseteq G_{\mathrm{con}} \subseteq G.$$

Then

G/G_{con} is discrete

$G_{\mathrm{con}}/G_{\mathrm{sol}}$ is a central extension of a product of simple connected Lie groups.

$G_{\mathrm{sol}}/G_{\mathrm{nil}}$ is abelian. A connected abelian Lie group is isomorphic to a product of copies of **R** and the circle group S^1.

$G_{\mathrm{nil}}/1$ is nilpotent, and therefore its ascending central series has all quotients abelian.

This can be used to reduce some problems about Lie groups (such as finding their unitary representations) to the same problems for connected simple groups and nilpotent and solvable subgroups of smaller dimension.

- The diffeomorphism group of a Lie group acts transitively on the Lie group

- Every Lie group is parallelizable, and hence an orientable manifold (there is a bundle isomorphism between its tangent bundle and the product of itself with the tangent space at the identity)

3.7   Infinite-dimensional Lie groups

Lie groups are often defined to be finite-dimensional, but there are many groups that resemble Lie groups, except for being infinite-dimensional. The simplest way to define infinite-dimensional Lie groups is to model them on Banach spaces, and in this case much of the basic theory is similar to that of finite-dimensional Lie groups. However this is inadequate for many applications, because many natural examples of infinite-dimensional Lie groups are not Banach manifolds. Instead one needs to define Lie groups modeled on more general locally convex topological vector spaces. In this case the relation between the Lie algebra and the Lie group becomes rather subtle, and several results about finite-dimensional Lie groups no longer hold.

The literature is not entirely uniform in its terminology as to exactly which properties of infinite-dimensional groups qualify the group for the prefix *Lie* in *Lie group*. On the Lie algebra side of affairs, things are simpler since the qualifying criteria for the prefix *Lie* in *Lie algebra* are purely algebraic. For example, an infinite-dimensional Lie algebra may or may not have a corresponding Lie group. That is, there may be a group corresponding to the Lie algebra, but it might not be nice enough to be called a Lie group, or the connection between the group and the Lie algebra might not be nice enough (e.g failure of the exponential map to be onto a neighborhood of the identity). It is the "nice enough" that is not universally defined.

Some of the examples that have been studied include:

- The group of diffeomorphisms of a manifold. Quite a lot is known about the group of diffeomorphisms of the circle. Its Lie algebra is (more or less) the Witt algebra, which has a central extension called the Virasoro algebra, used in string theory and conformal field theory. Diffeomorphism groups of compact manifolds of larger dimension are regular Fréchet Lie groups; very little about their structure is known.

The diffeomorphism group of spacetime sometimes appears in attempts to quantize gravity.

- The group of smooth maps from a manifold to a finite-dimensional Lie group is an example of a gauge group (with operation of pointwise multiplication), and is used in quantum field theory and Donaldson theory. If the manifold is a circle these are called loop groups, and have central extensions whose Lie algebras are (more or less) Kac–Moody algebras.

- There are infinite-dimensional analogues of general linear groups, orthogonal groups, and so on. One important aspect is that these may have *simpler* topological properties: see for example Kuiper's theorem. In M-Theory theory, for example, a 10 dimensional SU(N) gauge theory becomes an 11 dimensional theory when N becomes infinite.

- A specific example is that $SU(\infty)$ is equal to the group of area preserving diffeomorphisms of a torus.

3.8 See also

- Lie subgroup

- E_8

- Adjoint representation of a Lie group

- Adjoint endomorphism

- Haar measure

- Homogeneous space

- List of Lie group topics

- List of simple Lie groups

- Moufang polygon

- Riemannian manifold

- Representations of Lie groups

- Table of Lie groups

- Lie algebra

- Symmetry in quantum mechanics

- Lie group action

3.9   Notes

3.9.1   Explanatory notes

[1] having derivatives of all orders

[2] http://www.math.sunysb.edu/~{}vkiritch/MAT552/ProblemSet1.pdf

3.9.2   Citations

[1] Arthur Tresse (1893). "Sur les invariants différentiels des groupes continus de transformations". *Acta Mathematica* **18**: 1–88. doi:10.1007/bf02418270.

[2] Borel (2001).

[3] Helgason, Sigurdur (1978). *Differential Geometry, Lie Groups, and Symmetric Spaces*. New York: Academic Press. p. 131. ISBN 0-12-338460-5.

3.10   References

- Adams, John Frank (1969), *Lectures on Lie Groups*, Chicago Lectures in Mathematics, Chicago: Univ. of Chicago Press, ISBN 0-226-00527-5, MR 0252560.

- Borel, Armand (2001), *Essays in the history of Lie groups and algebraic groups*, History of Mathematics **21**, Providence, R.I.: American Mathematical Society, ISBN 978-0-8218-0288-5, MR 1847105

- Bourbaki, Nicolas, *Elements of mathematics: Lie groups and Lie algebras.* Chapters 1–3 ISBN 3-540-64242-0, Chapters 4–6 ISBN 3-540-42650-7, Chapters 7–9 ISBN 3-540-43405-4

- Chevalley, Claude (1946), *Theory of Lie groups*, Princeton: Princeton University Press, ISBN 0-691-04990-4.

- P. M. Cohn (1957) *Lie Groups*, Cambridge Tracts in Mathematical Physics.

- J. L. Coolidge (1940) *A History of Geometrical Methods*, pp 304–17, Oxford University Press (Dover Publications 2003).

- Fulton, William; Harris, Joe (1991), *Representation theory. A first course*, Graduate Texts in Mathematics, Readings in Mathematics **129**, New York: Springer-Verlag, ISBN 978-0-387-97495-8, MR 1153249, ISBN 978-0-387-97527-6

- Robert Gilmore (2008) *Lie groups, physics, and geometry: an introduction for physicists, engineers and chemists*, Cambridge University Press ISBN 9780521884006 .

- Hall, Brian C. (2003), *Lie Groups, Lie Algebras, and Representations: An Elementary Introduction*, Springer, ISBN 0-387-40122-9.

- F. Reese Harvey (1990) *Spinors and calibrations*, Academic Press, ISBN 0-12-329650-1 .

- Hawkins, Thomas (2000), *Emergence of the theory of Lie groups*, Sources and Studies in the History of Mathematics and Physical Sciences, Berlin, New York: Springer-Verlag, ISBN 978-0-387-98963-1, MR 1771134 Borel's review

- Helgason, Sigurdur (2001), *Differential geometry, Lie groups, and symmetric spaces*, Graduate Studies in Mathematics **34**, Providence, R.I.: American Mathematical Society, ISBN 978-0-8218-2848-9, MR 1834454

- Knapp, Anthony W. (2002), *Lie Groups Beyond an Introduction*, Progress in Mathematics **140** (2nd ed.), Boston: Birkhäuser, ISBN 0-8176-4259-5.

- Nijenhuis, Albert (1959). "Review: *Lie groups*, by P. M. Cohn". *Bulletin of the American Mathematical Society* **65** (6): 338–341. doi:10.1090/s0002-9904-1959-10358-x.

- Rossmann, Wulf (2001), *Lie Groups: An Introduction Through Linear Groups*, Oxford Graduate Texts in Mathematics, Oxford University Press, ISBN 978-0-19-859683-7. The 2003 reprint corrects several typographical mistakes.

- Sattinger, David H.; Weaver, O. L. (1986). *Lie groups and algebras with applications to physics, geometry, and mechanics*. Springer-Verlag. ISBN 3-540-96240-9. MR 0835009.

- Serre, Jean-Pierre (1965), *Lie Algebras and Lie Groups: 1964 Lectures given at Harvard University*, Lecture notes in mathematics **1500**, Springer, ISBN 3-540-55008-9.

- Stillwell, John (2008). *Naive Lie Theory*. Springer. ISBN 0-387-98289-2.

- Heldermann Verlag Journal of Lie Theory

- Warner, Frank W. (1983), *Foundations of differentiable manifolds and Lie groups*, Graduate Texts in Mathematics **94**, New York Berlin Heidelberg: Springer-Verlag, ISBN 978-0-387-90894-6, MR 0722297

- Steeb, Willi-Hans (2007), *Continuous Symmetries, Lie algebras, Differential Equations and Computer Algebra: second edition*, World Scientific Publishing, ISBN 981-270-809-X, MR 2382250.

- Lie Groups. Representation Theory and Symmetric Spaces Wolfgang Ziller, Vorlesung 2010

Chapter 4

Classical group

For the book by Weyl, see The Classical Groups.

In mathematics, the **classical groups** are defined as the special linear groups over the reals **R**, the complex numbers **C** and the quaternions **H** together with special[1] automorphism groups of symmetric or skew-symmetric bilinear forms and Hermitian or skew-Hermitian sesquilinear forms defined on real, complex and quaternionic finite-dimensional vector spaces.[2] Of these, the **complex classical Lie groups** are four infinite families of Lie groups that together with the exceptional groups exhaust the classification of simple Lie groups. The **compact classical groups** are compact real forms of the complex classical groups. The finite analogues of the classical groups are the **classical groups of Lie type**. The term "classical group" was coined by Hermann Weyl, it being the title of his 1939 monograph *The Classical Groups*.[3]

The classical groups form the deepest and most useful part of the subject of linear Lie groups.[4] Most types of classical groups find application in classical and modern physics. A few examples are the following. The rotation group SO(3) is a symmetry of Euclidean space and all fundamental laws of physics, the Lorentz group O(3,1) is a symmetry group of spacetime of special relativity. The special unitary group SU(3) is the symmetry group of quantum chromodynamics and the symplectic group Sp(m) finds application in hamiltonian mechanics and quantum mechanical versions of it.

4.1 The classical groups

The **classical groups** are exactly the general linear groups over **R**, **C** and **H** together with the automorphism groups of non-degenerate forms discussed below.[5] These groups are usually additionally restricted to the subgroups whose elements have determinant 1. The classical groups, with the determinant 1 condition, are listed in the table below. In the sequel, the determinant 1 condition is *not* used consistently in the interest of greater generality.

The **complex classical groups** are SL(n, **C**), SO(n, **C**) and Sp(m, **C**). A group is complex according to whether its Lie algebra is complex. The **real classical groups** refers to all of the classical groups since any Lie algebra is a real algebra. The **compact classical groups** are the compact real forms of the complex classical groups. These are, in turn, SU(n), SO(n) and Sp(m). One characterization of the compact real form is in terms of the Lie algebra **g**. If **g** = **u** + i**u**, the complexification of **u**, then if the connected group K generated by exp(X): $X \in$ **u** is a compact, K is a compact real form.[6]

The classical groups can uniformly be characterized in a different way using real forms. The classical groups (here with the determinant 1 condition, but this is not necessary) are the following:

The complex linear algebraic groups SL(n, **C**), SO(n, **C**), and Sp(n, **C**) together with their real forms.[7]

For instance, SO*($2n$) is a real form of SO($2n$, **C**), SU(p, q) is a real form of Sl(n, **C**), and Sl(n, **H**) is a real form of SO($2n$, **C**). Without the determinant 1 condition, replace the special linear groups with the corresponding general linear groups in the characterization. The algebraic groups in question are Lie groups, but the "algebraic" qualifier is needed to get the right notion of "real form".

4.2 Bilinear and sesquilinear forms

Main articles: Bilinear form and Sesquilinear form

The classical groups are defined in terms of forms defined on $\mathbf{R}^n$, $\mathbf{C}^n$, and $\mathbf{H}^n$, where $\mathbf{R}$ and $\mathbf{C}$ are the fields of the real and complex numbers. The quaternions, $\mathbf{H}$, do not constitute a field because multiplication does not commute; they form a division ring or a **skew field** or **non-commutative field**. However, it is still possible to define matrix quaternionic groups. For this reason, a vector space V is allowed to be defined over $\mathbf{R}$, $\mathbf{C}$, as well as $\mathbf{H}$ below. In the case of $\mathbf{H}$, V is a *right* vector space to make possible the representation of the group action as matrix multiplication from the *left*, just as for $\mathbf{R}$ and $\mathbf{C}$.[8]

A form $\varphi\colon V \times V \to F$ on some finite-dimensional right vector space over $F = \mathbf{R}$, $\mathbf{C}$, or $\mathbf{H}$ is bilinear if

$$\varphi(x\alpha, y\beta) = \alpha\varphi(x,y)\beta, \quad \forall x, y \in V, \forall \alpha, \beta \in F.$$

It is called **sesquilinear** if

$$\varphi(x\alpha, y\beta) = \bar{\alpha}\varphi(x,y)\beta, \quad \forall x, y \in V, \forall \alpha, \beta \in F.$$

These conventions are chosen because they work in all cases considered. An automorphism of φ is a map A in the set of linear operators on V such that

The set of all automorphisms of φ form a group, it is called the automorphism group of φ, denoted $\mathrm{Aut}(\varphi)$. This leads to a preliminary definition of a classical group:

> *A classical group is a group that preserves a bilinear or sesquilinear form on finite-dimensional vector spaces over* $\mathbf{R}$, $\mathbf{C}$ *or* $\mathbf{H}$.

This definition has shortcomings because there is some unnecessary redundancy. In the case of $F = \mathbf{R}$, bilinear is equivalent to sesquilinear. In the case of $F = \mathbf{H}$, there are no non-zero bilinear forms.[9]

4.2.1 Symmetric, skew-symmetric, Hermitian, and skew-Hermitian forms

A form is **symmetric** if

$$\varphi(x,y) = \varphi(y,x).$$

It is **skew-symmetric** if

$$\varphi(x,y) = -\varphi(y,x).$$

It is **Hermitian** if

$$\varphi(x, y) = \overline{\varphi(y, x)}$$

Finally, it is **skew-Hermitian** if

$$\varphi(x, y) = -\overline{\varphi(y, x)}.$$

A bilinear form φ is uniquely a sum of a symmetric form and a skew-symmetric form. A transformation preserving φ preserves both parts separately. The groups preserving symmetric and skew-symmetric forms can thus be studied separately. The same applies, mutatis mutandis, to Hermitian and skew-Hermitian forms. For this reason, for the purposes of classification, only purely symmetric, skew-symmetric, Hermitian, or skew-Hermitian forms are considered. The **normal forms** of the forms correspond to specific suitable choices of bases. These are bases giving the following normal forms in coordinates:

basis: (pseudo-)orthonormal in form symmetric Bilinear	$\varphi(x, y) = \pm\xi_1\eta_1 \pm \xi_2\eta_2 \pm \cdots \pm \xi_n\eta_n,$ $(\mathbf{R})$
basis: orthonormal in form symmetric Bilinear	$\varphi(x, y) = \xi_1\eta_1 + \xi_2\eta_2 + \cdots + \xi_n\eta_n,$ $(\mathbf{C})$
basis: symplectic in skew-symmetric Bilinear	$\varphi(x, y) = \xi_1\eta_{m+1} + \xi_2\eta_{m+2} + \cdots + \xi_m\eta_{2m=n}$
	$\qquad - \xi_{m+1}\eta_1 - \xi_{m+2}\eta_2 - \cdots - \xi_{2m=n}\eta_m,$ $(\mathbf{R}, \mathbf{C})$
Hermitian: Sesquilinear	$\varphi(x, y) = \pm\bar{\xi}_1\eta_1 \pm \bar{\xi}_2\eta_2 \pm \cdots \pm \bar{\xi}_n\eta_n,$ $(\mathbf{C}, \mathbf{H})$
skew-Hermitian: Sesquilinear	$\varphi(x, y) = \bar{\xi}_1\mathbf{j}\eta_1 + \bar{\xi}_2\mathbf{j}\eta_2 + \cdots + \bar{\xi}_n\mathbf{j}\eta_n,$ $(\mathbf{H}).$

The $\mathbf{j}$ in the skew-Hermitian form is the third basis element in the basis $(\mathbf{1}, \mathbf{i}, \mathbf{j}, \mathbf{k})$ for $\mathbf{H}$. Proof of existence of these bases and Sylvester's law of inertia, the independence of the number of plus- and minus-signs, p and q, in the symmetric and Hermitian forms, as well as the presence or absence of the fields in each expression, can be found in Rossmann (2002) or Goodman & Wallach (2009). The pair (p, q), and sometimes $p - q$, is called the **signature** of the form.

Explanation of occurrence of the fields R, C, H: There are no nontrivial bilinear forms over H. In the symmetric bilinear case, only forms over $\mathbf{R}$ have a signature. In other words, a complex bilinear form with "signature" (p, q) can, by a change of basis, be reduced to a form where all signs are "+" in the above expression, whereas this is impossible in the real case, in which $p - q$ is independent of the basis when put into this form. However, Hermitian forms have basis-independent signature in both the complex and the quaternionic case. (The real case reduces to the symmetric case.) A skew-Hermitian form on a complex vector space is rendered Hermitian by multiplication by i, so in this case, only $\mathbf{H}$ is interesting.

4.3 Automorphism groups

The first section presents the general framework. The other sections exhaust the qualitatively different cases that arise as automorphism groups of bilinear and sesquilinear forms on finite-dimensional vector spaces over $\mathbf{R}$, $\mathbf{C}$ and $\mathbf{H}$.

4.3.1 Aut(φ) – the automorphism group

Assume that φ is a non-degenerate form on a finite-dimensional vector space V over $\mathbf{R}$, $\mathbf{C}$ or $\mathbf{H}$. The automorphism group is defined, based on condition (**1**), as

$$\mathrm{Aut}(\varphi) = \{A \in \mathrm{GL}(V) : \varphi(Au, Av) = \varphi(x, y), \quad \forall x, y \in V\}.$$

Every $A \in Mn(V)$ has an adjoint A^φ with respect to φ defined by

Hermann Weyl, the author of The Classical Groups. Weyl made substantial contributions to the representation theory of the classical groups.

Using this definition in condition (**1**), the automorphism group is seen to be given by

Fix a basis for *V*. In terms of this basis, put

$$\varphi(x,y) = \sum \xi_i \varphi_{ij} \eta_j$$

where ξi, ηj are the components of x, y. This is appropriate for the bilinear forms. Sesquilinear forms have similar expressions and are treated separately later. In matrix notation one finds

$$\varphi(x,y) = x^T \Phi y$$

and

from (2) where Φ is the matrix (φij). The non-degeneracy condition means precisely that Φ is invertible, so the adjoint always exists. Aut(φ) expressed with this becomes

$$\mathrm{Aut}(\varphi) = \{A \in \mathrm{GL}(V) : \Phi^{-1} A^T \Phi A = 1\}.$$

The Lie algebra **aut**(φ) of the automorphism groups can be written down immediately. Abstractly, $X \in$ **aut**(φ) if and only if

$$(e^{tX})^\varphi e^{tX} = 1$$

for all t, corresponding to the condition in (3) under the exponential mapping of Lie algebras, so that

$$\mathfrak{aut}(\varphi) = \{X \in M_n(V) : X^\varphi = -X\},$$

or in a basis

as is seen using the power series expansion of the exponential mapping and the linearity of the involved operations. Conversely, suppose that $X \in$ **aut**(φ). Then, using the above result, $\varphi(Xx, y) = \varphi(x, X^\varphi y) = -\varphi(x, Xy)$. Thus the Lie algebra can be characterized without reference to a basis, or the adjoint, as

$$\mathfrak{aut}(\varphi) = \{X \in M_n(V) : \varphi(Xx, y) = -\varphi(x, Xy), \quad \forall x, y \in V\}.$$

The normal form for φ will be given for each classical group below. From that normal form, the matrix Φ can be read off directly. Consequently, expressions for the adjoint and the Lie algebras can be obtained using formulas (4) and (5). This is demonstrated below in most of the non-trivial cases.

4.3.2   Bilinear case

When the form is symmetric, Aut(φ) is called O(φ). When it is skew-symmetric then Aut(φ) is called Sp(φ). This applies to the real and the complex cases. The quaternionic case is empty since no nonzero bilinear forms exists on quaternionic vector spaces.[12]

Real case

The real case breaks up into two cases, the symmetric and the antisymmetric forms that should be treated separately.

O(p, q) and O(n) – the orthogonal groups   Main articles: Orthogonal group and Indefinite orthogonal group

If φ is symmetric and the vector space is real, a basis may be chosen so that

$$\varphi(x, y) = \pm \xi_1 \eta_1 \pm \xi_1 \eta_1 \cdots \pm \xi_n \eta_n.$$

The number of plus and minus-signs are independent of the particular basis.[13] In the case $V = \mathbf{R}^n$ one writes O(φ) = O(p, q) where p is the number of plus signs and q is the number of minus-signs, $p + q = n$. If $q = 0$ the notation is O(n). The matrix Φ is in this case

$$\Phi = \begin{pmatrix} I_p & 0 \\ 0 & -I_q \end{pmatrix} \equiv I_{p,q}$$

after reordering the basis if necessary. The adjoint operation (**4**) then becomes

$$A^\varphi = \begin{pmatrix} I_p & 0 \\ 0 & -I_q \end{pmatrix} \begin{pmatrix} A_{11} & \cdots \\ \cdots & A_{nn} \end{pmatrix}^{\mathrm{T}} \begin{pmatrix} I_p & 0 \\ 0 & -I_q \end{pmatrix},$$

which reduces to the usual transpose when p or q is 0. The Lie algebra is found using equation (**5**) and a suitable ansatz (this is detailed for the case of Sp(m, **R** below),

$$\mathfrak{o}(p, q) = \left\{ \begin{pmatrix} X_{p \times p} & Y_{p \times q} \\ Y^{\mathrm{T}} & W_{q \times q} \end{pmatrix} \middle| X^{\mathrm{T}} = -X, \quad W^{\mathrm{T}} = -W \right\},$$

and the group according to (**3**) is given by

$$O(p, q) = \{ g \in \mathrm{GL}(n, \mathbb{R}) | I_{p,q}^{-1} g^{\mathrm{T}} I_{p,q} g = I \}.$$

The groups O(p, q) and O(q, p) are isomorphic through the map

$$O(p, q) \to O(q, p), \quad g \to \sigma g \sigma^{-1}, \quad \sigma = \begin{bmatrix} 0 & 0 & \cdots & 1 \\ \vdots & \vdots & \ddots & \vdots \\ 0 & 1 & \cdots & 0 \\ 1 & 0 & \cdots & 0 \end{bmatrix}.$$

For example, the Lie algebra of the Lorentz group could be written as

$$\mathfrak{o}(3, 1) = \mathrm{span} \left\{ \begin{pmatrix} 0 & 1 & 0 & 0 \\ -1 & 0 & 0 & 0 \\ 0 & 0 & 0 & 0 \\ 0 & 0 & 0 & 0 \end{pmatrix}, \begin{pmatrix} 0 & 0 & -1 & 0 \\ 0 & 0 & 0 & 0 \\ 1 & 0 & 0 & 0 \\ 0 & 0 & 0 & 0 \end{pmatrix}, \begin{pmatrix} 0 & 0 & 0 & 0 \\ 0 & 0 & 1 & 0 \\ 0 & -1 & 0 & 0 \\ 0 & 0 & 0 & 0 \end{pmatrix}, \begin{pmatrix} 0 & 0 & 0 & 1 \\ 0 & 0 & 0 & 0 \\ 0 & 0 & 0 & 0 \\ 1 & 0 & 0 & 0 \end{pmatrix}, \begin{pmatrix} 0 & 0 & 0 & 0 \\ 0 & 0 & 0 & 1 \\ 0 & 0 & 0 & 0 \\ 0 & 1 & 0 & 0 \end{pmatrix}, \begin{pmatrix} 0 & 0 & 0 & 0 \\ 0 & 0 & 0 & 0 \\ 0 & 0 & 0 & 1 \\ 0 & 0 & 1 & 0 \end{pmatrix} \right\}.$$

Naturally, it is possible to rearrange so that the q-block is the upper left (or any other block). Here the "time component" end up as the fourth coordinate in a physical interpretation, and not the first as may be more common.

Sp(m, R) – the real symplectic group   Main article: Symplectic group

If φ is skew-symmetric and the vector space is real, there is a basis giving

$$\varphi(x, y) = \xi_1 \eta_{m+1} + \xi_2 \eta_{m+2} \cdots + \xi_m \eta_{2m=n} - \xi_{m+1} \eta_1 - \xi_{m+2} \eta_2 \cdots - \xi_{2m=n} \eta_m,$$

where $n = 2m$. For $\text{Aut}(\varphi)$ one writes $\text{Sp}(\varphi) = \text{Sp}(V)$ In case $V = \mathbf{R}^n = \mathbf{R}^{2m}$ one writes $\text{Sp}(m, \mathbf{R})$ or $\text{Sp}(2m, \mathbf{R})$. From the normal form one reads off

$$\Phi = \begin{pmatrix} 0_m & I_m \\ -I_m & 0_m \end{pmatrix} = J_m.$$

By making the ansatz

$$V = \begin{pmatrix} X & Y \\ Z & W \end{pmatrix},$$

where X, Y, Z, W are m-dimensional matrices and considering (5),

$$\begin{pmatrix} 0_m & -I_m \\ I_m & 0_m \end{pmatrix} \begin{pmatrix} X & Y \\ Z & W \end{pmatrix}^{\mathrm{T}} \begin{pmatrix} 0_m & I_m \\ -I_m & 0_m \end{pmatrix} = - \begin{pmatrix} X & Y \\ Z & W \end{pmatrix}$$

one finds the Lie algebra of $\text{Sp}(m, \mathbf{R})$,

$$\mathfrak{sp}(m, \mathbb{R}) = \left\{ X \in M_n(\mathbb{R}) : J_m X + X^{\mathrm{T}} J_m = 0 \right\} = \left\{ \begin{pmatrix} X & Y \\ Z & -X^{\mathrm{T}} \end{pmatrix} \middle| Y^{\mathrm{T}} = Y, Z^{\mathrm{T}} = Z \right\},$$

and the group is given by

$$\text{Sp}(m, \mathbb{R}) = \{ g \in M_n(\mathbb{R}) | g^{\mathrm{T}} J_m g = J_m \}.$$

Complex case

Like in the real case, there are two cases, the symmetric and the antisymmetric case that each yield a family of classical groups.

O(n, C) – the complex orthogonal group   Main article: Complex orthogonal group

If case φ is symmetric and the vector space is complex, a basis

$$\varphi(x, y) = \xi_1 \eta_1 + \xi_1 \eta_1 \cdots + \xi_n \eta_n$$

with only plus-signs can be used. The automorphism group is in the case of $V = \mathbf{C}^n$ called O(n, **C**). The lie algebra is simply a special case of that for $\mathbf{o}(p, q)$,

$$\mathfrak{o}(n, \mathbb{C}) = \mathfrak{so}(n, \mathbb{C}) = \{ X | X^{\mathrm{T}} = -X \},$$

and the group is given by

$$\text{O}(n, \mathbb{C}) = \{ g | g^{\mathrm{T}} g = I_n \}.$$

In terms of classification of simple Lie algebras, the **so**(n) are split into two classes, those with n odd with root system Bn and n even with root system Dn.

Sp(m, C) – the complex symplectic group   Main article: Symplectic group

For φ skew-symmetric and the vector space complex, the same formula,

$$\varphi(x,y) = \xi_1\eta_{m+1} + \xi_2\eta_{m+2}\cdots + \xi_m\eta_{2m=n} - \xi_{m+1}\eta_1 - \xi_{m+2}\eta_2\cdots - \xi_{2m=n}\eta_m,$$

applies as in the real case. For Aut(φ) one writes Sp(φ) = Sp(V) In case $V = \mathbb{C}^n = \mathbb{C}^{2m}$ one writes Sp(m, $\mathbb{C}$) or Sp($2m$, $\mathbb{C}$). The Lie algebra parallels that of **sp**(m, $\mathbb{R}$),

$$\mathfrak{sp}(m,\mathbb{C}) = \left\{X \in M_n(\mathbb{C}) : J_m X + X^\mathsf{T} J_m = 0\right\} = \left\{\begin{pmatrix} X & Y \\ Z & -X^\mathsf{T} \end{pmatrix}\middle| Y^\mathsf{T} = Y, Z^\mathsf{T} = Z\right\},$$

and the group is given by

$$\mathrm{Sp}(m,\mathbb{C}) = \{g \in M_n(\mathbb{C})|g^\mathsf{T} J_m g = J_m\}.$$

4.3.3   Sesquilinear case

In the sequilinear case, one makes a slightly different ansatz for the form in terms of a basis,

$$\varphi(x,y) = \sum \bar{\xi}_i \varphi_{ij} \eta_j.$$

The other expressions that get modified are

$$\varphi(x,y) = x^*\Phi y, \qquad A^\varphi = \Phi^{-1} A^* \Phi, \text{ [14]}$$
$$\mathrm{Aut}(\varphi) = \{A \in \mathrm{GL}(V) : \Phi^{-1} A^* \Phi A = 1\},$$

The real case, of course, provides nothing new. The complex and the quaternionic case will be considered below.

Complex case

From a qualitative point of view, consideration of skew-Hermitean forms (up to isomorphism) provide no new groups; multiplication by i renders a skew-Hermitean form Hermitean, and vice versa. Thus only the Hermitian case needs to be considered.

U(p, q) and U(n) – the unitary groups   Main article: Unitary group

A non-degenerate hermitian form has the normal form

$$\varphi(x,y) = \pm\bar{\xi}_1\eta_1 \pm \bar{\xi}_2\eta_2 \cdots \pm \bar{\xi}_n\eta_n.$$

As in the bilinear case, the signature (p, q) is independent of the basis. The automorphism group is denoted U(V), or, in the case of $V = \mathbb{C}^n$, U(p, q). If $q = 0$ the notation is U(n). In this case, Φ takes the form

$$\Phi = \begin{pmatrix} 1_p & 0 \\ 0 & -1_q \end{pmatrix} = I_{p,q},$$

and the Lie algebra is given by

$$\mathfrak{u}(p,q) = \left\{ \begin{pmatrix} X_{p\times p} & Z_{p\times q} \\ \overline{Z}^{\mathrm{T}} & Y_{q\times q} \end{pmatrix} \middle| \overline{X}^{\mathrm{T}} = -X, \quad \overline{Y}^{\mathrm{T}} = -Y \right\}.$$

The group is given by

$$\mathrm{U}(p,q) = \{ g | I_{p,q}^{-1} g^* I_{p,q} g = I \}.$$

Quaternionic case

The space $\mathbf{H}^n$ is considered as a *right* vector space over $\mathbf{H}$. This way, $A(vh) = (Av)h$ for a quaternion h, a quaternion column vector v and quaternion matrix A. If $\mathbf{H}^n$ was a *left* vector space over $\mathbf{H}$, then matrix multiplication from the *right* on row vectors would be required to maintain linearity. This does not correspond to the usual linear operation of a group on a vector space when a basis is given, which is matrix multiplication from the *left* on column vectors. Thus V is henceforth a right vector space over $\mathbf{H}$. Even so, care must be taken due to the non-commutative nature of $\mathbf{H}$. The (mostly obvious) details are skipped because complex representations will be used.

When dealing with quaternionic groups it is convenient to represent quaternions using complex 2×2-matrices,

With this representation, quaternionic multiplication becomes matrix multiplication and quaternionic conjugation becomes taking the Hermitian adjoint. Moreover, if a quaternion according to the complex encoding $q = x + \mathbf{j}y$ is given as a column vector $(x, y)^{\mathrm{T}}$, then multiplication from the left by a matrix representation of a quaternion produces a new column vector representing the correct quaternion. This representation differs slightly from a more common representation found in the quaternion article. The more common convention would force multiplication from the right on a row matrix to achieve the same thing.

Incidentally, the representation above makes it clear that the group of unit quaternions ($\alpha\overline{\alpha} + \beta\overline{\beta} = 1 = \det Q$) is isomorphic to SU(2).

Quaternionic $n\times n$-matrices matrices can, by obvious extension, be represented by $2n\times 2n$ block-matrices of complex numbers.[16] If one agrees to represent a quaternionic $n\times 1$ column vector by a $2n\times 1$ column vector with complex numbers according to the encoding of above, with the upper n numbers being the αi and the lower n the βi, then a quaternionic $n\times n$-matrix becomes a complex $2n\times 2n$-matrix exactly of the form given above, but now with α and β $n\times n$-matrices. More formally

A matrix $T \in \mathrm{GL}(2n, \mathbf{C})$ has the form displayed in (**8**) if and only if $JnT = TJn$. With these identifications,

$$\mathbb{H}^n \approx \mathbb{C}^{2n}, M_n(\mathbb{H}) \approx \left\{ T \in M_{2n}(\mathbb{C}) \middle| J_n T = \overline{T} J_n, \quad J_n = \begin{pmatrix} 0 & I_n \\ -I_n & 0 \end{pmatrix} \right\}.$$

The space $Mn(\mathbf{H}) \subset M_{2n}(\mathbf{C})$ is a real algebra, but it is not a complex subspace of $M_{2n}(\mathbf{C})$. Multiplication (from the left) by $\mathbf{i}$ in $Mn(\mathbf{H})$ using entry-wise quaternionic multiplication and then mapping to the image in $M_{2n}(\mathbf{C})$ yields a different result than multiplying entry-wise by i directly in $M_2n(\mathbf{C})$. The quaternionic multiplication rules give $\mathbf{i}(X + \mathbf{j}Y) = (\mathbf{i}X) + \mathbf{j}(-\mathbf{i}Y)$ where the new X and Y are inside the parentheses.

The action of the quaternionic matrices on quaternionic vectors is now represented by complex quantities, but otherwise it is the same as for "ordinary" matrices and vectors. The quaternionic groups are thus embedded in $M_2n(C)$ where n is the dimension of the quaternionic matrices.

The determinant of a quaternionic matrix is defined in this representation as being the ordinary complex determinant of its representative matrix. The non-commutative nature of quaternionic multiplication would, in the quaternionic representation of matrices, be ambiguous. The way $Mn(\mathbf{H})$ is embedded in $M_2n(\mathbf{C})$ is not unique, but all such embeddings are related through $g \mapsto AgA^{-1}$, $g \in \mathrm{GL}(2n, \mathbf{C})$ for $A \in \mathrm{O}(2n, \mathbf{C})$, leaving the determinant unaffected.[17] The name of $\mathrm{SL}(n, \mathbf{H})$ in this complex guise is $\mathrm{SU}^*(2n)$.

As opposed to in the case of $\mathbf{C}$, both the Hermitian and the skew-Hermitean case bring in something new when $\mathbf{H}$ is considered, so these cases are considered separately.

GL(*n*, H) and SL(*n*, H)   Under the identification above,

$$\mathrm{GL}(n, \mathbb{H}) = \{g \in \mathrm{GL}(2n, \mathbb{C}) | Jg = \bar{g}J, \det\ g \neq 0\} \equiv \mathrm{U}^*(2n).$$

Its Lie algebra $\mathbf{gl}(n, \mathbf{H})$ is the set of all matrices in the image of the mapping $Mn(\mathbf{H}) \leftrightarrow M_2n(\mathbf{C})$ of above,

$$\mathfrak{gl}(n, \mathbb{H}) = \left\{ \begin{pmatrix} X & -\overline{Y} \\ Y & \overline{X} \end{pmatrix} \middle| X, Y \in \mathfrak{gl}(n, \mathbb{C}) \right\} \equiv \mathfrak{u}^*(2n).$$

The quaternionic special linear group is given by

$$\mathrm{SL}(n, \mathbb{H}) = \{g \in \mathrm{GL}(n, \mathbb{H}) | \det g = 1\} \equiv \mathrm{SU}^*(2n),$$

where the determinant is taken on the matrices in $\mathbf{C}^{2n}$. The Lie algebra is

$$\mathfrak{sl}(n, \mathbb{H}) = \left\{ \begin{pmatrix} X & -\overline{Y} \\ Y & \overline{X} \end{pmatrix} \middle| \mathrm{Tr}\, X = 0 \right\} \equiv \mathfrak{su}^*(2n).$$

Sp(*p*, *q*) – the quaternionic unitary group   As above in the complex case, the normal form is

$$\varphi(x, y) = \pm \bar{\xi}_1 \eta_1 \pm \bar{\xi}_2 \eta_2 \cdots \pm \bar{\xi}_n \eta_n$$

and the number of plus-signs is independent of basis. When $V = \mathbf{H}^n$ with this form, $\mathrm{Sp}(\varphi) = \mathrm{Sp}(p, q)$. The reason for the notation is that the group can be represented, using the above prescription, as a subgroup of $\mathrm{Sp}(n, \mathbf{C})$ preserving a complex-hermitian form of signature $(2p, 2q)$[18] If p or $q = 0$ the group is denoted $\mathrm{U}(n, \mathbf{H})$. It is sometimes called the **hyperunitary group**.

In quaternionic notation,

$$\Phi = \begin{pmatrix} I_p & 0 \\ 0 & -I_q \end{pmatrix} = I_{p,q}$$

meaning that *quaternionic* matrices of the form

will satisfy

$$\Phi^{-1}Q^*\Phi = -Q,$$

see the section about **u**(p, q). Caution needs to be exercised when dealing with quaternionic matrix multiplication, but here only I and $-I$ are involved and these commute with every quaternion matrix. Now apply prescription (**8**) to each block,

$$\mathcal{X} = \begin{pmatrix} X_{1(p\times p)} & -\overline{X}_2 \\ X_2 & \overline{X}_1 \end{pmatrix}, \mathcal{Y} = \begin{pmatrix} Y_{1(q\times q)} & -\overline{Y}_2 \\ Y_2 & \overline{Y}_1 \end{pmatrix}, \mathcal{Z} = \begin{pmatrix} Z_{1(p\times q)} & -\overline{Z}_2 \\ Z_2 & \overline{Z}_1 \end{pmatrix},$$

and the relations in (**9**) will be satisfied if

$$X_1^* = -X, Y_1^* = -Y.$$

The Lie algebra becomes

$$\mathfrak{sp}(p,q) = \left\{ \left(\begin{bmatrix} X_{1(p\times p)} & -\overline{X}_2 \\ X_2 & \overline{X}_1 \\ Z_{1(p\times q)} & -\overline{Z}_2 \\ Z_2 & \overline{Z}_1 \end{bmatrix}^* \begin{bmatrix} Z_{1(p\times q)} & -\overline{Z}_2 \\ Z_2 & \overline{Z}_1 \\ Y_{1(q\times q)} & -\overline{Y}_2 \\ Y_2 & \overline{Y}_1 \end{bmatrix} \right) \middle| X_1^* = -X, Y_1^* = -Y \right\}.$$

The group is given by

$$\mathrm{Sp}(p,q) = \{g \in \mathrm{GL}(n,\mathbb{H}) | I_{p,q}^{-1} g^* I_{p,q} g = I_{p+q}\} = \{g \in \mathrm{GL}(2n,\mathbb{C}) | K_{p,q}^{-1} g^* K_{p,q} g = I_{2(p+q)}, \qquad K = \mathrm{diag}(I_{p,q}, I_{p,q})\}.$$

Returning to the normal form of $\varphi(w,z)$ for Sp(p, q), make the substitutions $w \to u + jv$ and $z \to x + jy$ with u, v, x, y $\in$ **C**^{n}. Then

$$\varphi(w,z) = \begin{bmatrix} u^* & v^* \end{bmatrix} K_{p,q} \begin{bmatrix} x \\ y \end{bmatrix} + j \begin{bmatrix} u & -v \end{bmatrix} K_{p,q} \begin{bmatrix} y \\ x \end{bmatrix} = \varphi_1(w,z) + \mathbf{j}\varphi_2(w,z), \qquad K_{p,q} = \mathrm{diag}(I_{p,q}, I_{p,q})$$

viewed as a **H**-valued form on **C**^{2n}.[19] Thus the elements of Sp(p, q), viewed as linear transformations of **C**^{2n}, preserve both a Hermitian form of signature (2p, 2q)and a non-degenerate skew-symmetric form. Both forms take purely complex values and due to the prefactor of **j** of the second form, they are separately conserved. This means that

$$\mathrm{Sp}(p,q) = \mathrm{U}(\mathbb{C}^{2n}, \varphi_1) \cap \mathrm{Sp}(\mathbb{C}^{2n}, \varphi_2)$$

and this explains both the name of the group and the notation.

O*(2n)= O(n, H)- quaternionic orthogonal group   The normal form for a skew-hermitian form is given by

$$\varphi(x,y) = \bar{\xi}_1 \mathbf{j} \eta_1 + \bar{\xi}_2 \mathbf{j} \eta_2 \cdots + \bar{\xi}_n \mathbf{j} \eta_n,$$

where $\mathbf{j}$ is the third basis quaternion in the ordered listing $(\mathbf{1}, \mathbf{i}, \mathbf{j}, \mathbf{k})$. In this case, $\mathrm{Aut}(\varphi) = \mathrm{O}^*(2n)$ may be realized, using the complex matrix encoding of above, as a subgroup of $\mathrm{O}(2n, \mathbf{C})$ which preserves a non-degenerate complex skew-hermitian form of signature (n, n).[20] From the normal form one sees that in quaternionic notation

$$\Phi = \begin{pmatrix} \mathbf{j} & 0 & \cdots & 0 \\ 0 & \mathbf{j} & \cdots & \vdots \\ \vdots & & \ddots & \\ 0 & \cdots & 0 & \mathbf{j} \end{pmatrix} \equiv \mathbf{j}_n$$

and from (6) follows that

for $V \in \mathbf{o}(2n)$. Now put

$$V = X + \mathbf{j}Y \leftrightarrow \begin{pmatrix} X & -\overline{Y} \\ Y & \overline{X} \end{pmatrix}$$

according to prescription (8). The same prescription yields for Φ,

$$\Phi \leftrightarrow \begin{pmatrix} 0 & -I_n \\ I_n & 0 \end{pmatrix} \equiv J_n.$$

Now the last condition in (9) in complex notation reads

$$\begin{pmatrix} X & -\overline{Y} \\ Y & \overline{X} \end{pmatrix}^* = \begin{pmatrix} 0 & -I_n \\ I_n & 0 \end{pmatrix} \begin{pmatrix} X & -\overline{Y} \\ Y & \overline{X} \end{pmatrix} \begin{pmatrix} 0 & -I_n \\ I_n & 0 \end{pmatrix} \Leftrightarrow X^{\mathrm{T}} = -X, \quad \overline{Y}^{\mathrm{T}} = Y.$$

The Lie algebra becomes

$$\mathbf{o}^*(2n) = \left\{ \begin{pmatrix} X & -\overline{Y} \\ Y & \overline{X} \end{pmatrix} \middle| X^{\mathrm{T}} = -X, \quad \overline{Y}^{\mathrm{T}} = Y \right\},$$

and the group is given by

$$\mathrm{O}^*(2n) = \{ g \in \mathrm{GL}(n, \mathbb{H}) | \mathbf{j}_n^{-1} g^* \mathbf{j}_n g = I_n \} = \{ g \in \mathrm{GL}(2n, \mathbb{C}) | J_n^{-1} g^* J_n g = I_{2n} \}.$$

The group $\mathrm{SO}^*(2n)$ can be characterized as

$$\mathrm{O}^*(2n) = \{ g \in \mathrm{O}(2n, \mathbb{C}) | \theta(\overline{g}) = g \}, \text{ [21]}$$

where the map $\theta \colon \mathrm{GL}(2n, \mathbf{C}) \to \mathrm{GL}(2n, \mathbf{C})$ is defined by $g \mapsto -J_{2n} g J_{2n}$. Also, the form determining the group can be viewed as a $\mathbf{H}$-valued form on $\mathbf{C}^{2n}$.[22] Make the substitutions $x \to w_1 + iw_2$ and $y \to z_1 + iz_2$ in the expression for the form. Then

$$\varphi(x, y) = \overline{w}_2 I_n z_1 - \overline{w}_1 I_n z_2 + \mathbf{j}(w_1 I_n z_1 + w_2 I_n z_2) = \overline{\varphi_1(w, z)} + \mathbf{j}\varphi_2(w, z).$$

The form φ_1 is Hermitian (while the first form on the left hand side is skew-Hermitian) of signature (n, n). The signature is made evident by a change of basis from $(\mathbf{e}, \mathbf{f})$ to $((\mathbf{e} + i\mathbf{f})/\sqrt{2}, (\mathbf{e} - i\mathbf{f})/\sqrt{2})$ where $\mathbf{e}, \mathbf{f}$ are the first and last n basis vectors respectively. The second form, φ_2 is symmetric positive definite. Thus, due to the factor $\mathbf{j}$, $\mathbf{O}^*(2n)$ preserves both separately and it may be concluded that

$$O^*(2n) = O(2n, \mathbb{C}) \cap U(\mathbb{C}^{2n}, \varphi_1),$$

and the notation "O" is explained.

4.4   Classical groups over general fields or algebras

Classical groups, more broadly considered in algebra, provide particularly interesting matrix groups. When the field F of coefficients of the matrix group is either real number or complex numbers, these groups are just the classical Lie groups. When the ground field is a finite field, then the classical groups are groups of Lie type. These groups play an important role in the classification of finite simple groups. Also, one may consider classical groups over a unital associative algebra R over F; where $R = \mathbf{H}$ (an algebra over reals) represents an important case. For the sake of generality the article will refer to groups over R, where R may be the ground field F itself.

Considering their abstract group theory, many linear groups have a "**special**" subgroup, usually consisting of the elements of determinant 1 over the ground field, and most of them have associated "**projective**" quotients, which are the quotients by the center of the group. For orthogonal groups in characteristic 2 "S" has a different meaning.

The word "**general**" in front of a group name usually means that the group is allowed to multiply some sort of form by a constant, rather than leaving it fixed. The subscript n usually indicates the dimension of the module on which the group is acting; it is a vector space if $R = F$. Caveat: this notation clashes somewhat with the n of Dynkin diagrams, which is the rank.

4.4.1   General and special linear groups

The general linear group $\mathrm{GL}n(R)$ is the group of all R-linear automorphisms of R^n. There is a subgroup: the special linear group $\mathrm{SL}n(R)$, and their quotients: the projective general linear group $\mathrm{PGL}n(R) = \mathrm{GL}n(R)/Z(\mathrm{GL}n(R))$ and the projective special linear group $\mathrm{PSL}n(R) = \mathrm{SL}n(R)/Z(\mathrm{SL}n(R))$. The projective special linear group $\mathrm{PSL}n(F)$ over a field F is simple for $n \geq 2$, except for the two cases when $n = 2$ and the field has order 2 or 3.

4.4.2   Unitary groups

The unitary group $\mathrm{U}n(R)$ is a group preserving a sesquilinear form on a module. There is a subgroup, the special unitary group $\mathrm{SU}n(R)$ and their quotients the projective unitary group $\mathrm{PU}n(R) = \mathrm{U}n(R)/Z(\mathrm{U}n(R))$ and the projective special unitary group $\mathrm{PSU}n(R) = \mathrm{SU}n(R)/Z(\mathrm{SU}n(R))$

4.4.3   Symplectic groups

The symplectic group $\mathrm{Sp}_2n(R)$ preserves a skew symmetric form on a module. It has a quotient, the projective symplectic group $\mathrm{PSp}_2n(R)$. The general symplectic group $\mathrm{GSp}_2n(R)$ consists of the automorphisms of a module multiplying a skew symmetric form by some invertible scalar. The projective symplectic group $\mathrm{PSp}_2n(\mathbf{F}q)$ over a finite field is simple for $n \geq 1$, except for the two cases when $n = 1$ and the field has order 2 or 3.

4.4.4   Orthogonal groups

The orthogonal group $\mathrm{O}n(R)$ preserves a non-degenerate quadratic form on a module. There is a subgroup, the special orthogonal group $\mathrm{SO}n(R)$ and quotients, the projective orthogonal group $\mathrm{PO}n(R)$, and the projective special orthogonal group PSO$n(R)$. In characteristic 2 the determinant is always 1, so the special orthogonal group is often defined as the subgroup of elements of Dickson invariant 1.

There is a nameless group often denoted by $\Omega n(R)$ consisting of the elements of the orthogonal group of elements of spinor norm 1, with corresponding subgroup and quotient groups S$\Omega n(R)$, P$\Omega n(R)$, PS$\Omega n(R)$. (For positive definite quadratic forms over the reals, the group Ω happens to be the same as the orthogonal group, but in general it is smaller.) There is also a double cover of $\Omega n(R)$, called the pin group Pin$n(R)$, and it has a subgroup called the spin group Spin$n(R)$. The general orthogonal group GO$n(R)$ consists of the automorphisms of a module multiplying a quadratic form by some invertible scalar.

4.4.5 Notational conventions

For more details on this topic, see Group of Lie type § Notation issues.

4.5 Contrast with exceptional Lie groups

Contrasting with the classical Lie groups are the exceptional Lie groups, G_2, F_4, E_6, E_7, E_8, which share their abstract properties, but not their familiarity.[23] These were only discovered around 1890 in the classification of the simple Lie algebras over the complex numbers by Wilhelm Killing and Élie Cartan.

4.6 Notes

[1] Here, *special* means the subgroup of the full automorphism group whose elements have determinant 1.

[2] Rossmann 2002 p. 94.

[3] Weyl 1939

[4] Rossmann 2002 p. 91.

[5] Rossmann 2002 p, 94

[6] Rossmann 2002 p. 103.

[7] Goodman & Wallach 2009 See end of chapter 1.

[8] Rossmann 2002p. 93.

[9] Rossmann 2002 p. 105

[10] Rossmann 2002 p. 91

[11] Rossmann 2002 p. 92

[12] Rossmann 2002 p. 105

[13] Rossmann 2002 p. 107.

[14] Rossmann 2002 p. 93

[15] Rossmann 2002 p. 95.

[16] Rossmann 2002 p. 94.

[17] Goodman & Wallach 2009 Exercise 14, Section 1.1.

[18] Rossmann 2002 p. 94.

[19] Goodman & Wallach 2009Exercise 11, Chapter 1.

[20] Rossmann 2002 p. 94.

[21] Goodman & Wallach 2009 p.11.

[22] Goodman & Wallach 2009 Exercise 12 Chapter 1.

[23] Wybourne, B. G. (1974). *Classical Groups for Physicists*, Wiley-Interscience. ISBN 0471965057 .

4.7   References

- E. Artin, *Geometric algebra*, Interscience (1957)

- Dieudonné, Jean (1955), *La géométrie des groupes classiques*, Ergebnisse der Mathematik und ihrer Grenzgebiete (N.F.), Heft 5, Berlin, New York: Springer-Verlag, ISBN 978-0-387-05391-2, MR 0072144

- Goodman, Roe; Wallach, Nolan R. (2009), *Symmetry, Representations,and Invariants*, Graduate texts in mathematics **255**, Springer-Verlag, ISBN 978-0-387-79851-6

- Knapp, A. W. (2002). *Lie groups beyond an introduction*. Progress in Mathematics **120** (2nd ed.). Boston·Basel·Berlin: Birkhäuser. ISBN 0-8176-4259-5.

- V. L. Popov (2001), "Classical group", in Hazewinkel, Michiel, *Encyclopedia of Mathematics*, Springer, ISBN 978-1-55608-010-4

- Rossmann, Wulf (2002), *Lie Groups - An Introduction Through Linear Groups*, Oxford Graduate Texts in Mathematics, Oxford Science Publications, ISBN 0 19 859683 9

Chapter 5

Simple Lie group

In group theory, a **simple Lie group** is a connected non-abelian Lie group G which does not have nontrivial connected normal subgroups.

A **simple Lie algebra** is a non-abelian Lie algebra whose only ideals are 0 and itself (or equivalently, a Lie algebra of dimension 2 or more, whose only ideals are 0 and itself). A direct sum of simple Lie algebras is called a semisimple Lie algebra.

An equivalent definition of a simple Lie group follows from the Lie correspondence: a connected Lie group is simple if its Lie algebra is simple. An important technical point is that a simple Lie group may contain *discrete* normal subgroups, hence being a simple Lie group is different from being simple as an abstract group.

Simple Lie groups include many classical Lie groups, which provide a group-theoretic underpinning for spherical geometry, projective geometry and related geometries in the sense of Felix Klein's Erlangen programme. It emerged in the course of classification of simple Lie groups that there exist also several exceptional possibilities not corresponding to any familiar geometry. These *exceptional groups* account for many special examples and configurations in other branches of mathematics, as well as contemporary theoretical physics.

While the notion of a simple Lie group is satisfying from the axiomatic perspective, in applications of Lie theory, such as the theory of Riemannian symmetric spaces, somewhat more general notions of semisimple and reductive Lie groups proved to be even more useful. In particular, every connected compact Lie group is reductive, and the study of representations of general reductive groups is a major branch of representation theory.

5.1   Comments on the definition

Unfortunately there is no single standard definition of a simple Lie group. The definition given above is sometimes varied in the following ways:

- Connectedness: Usually simple Lie groups are connected by definition. This excludes discrete simple groups (these are zero-dimensional Lie groups that are simple as abstract groups) as well as disconnected orthogonal groups.

- Center: Usually simple Lie groups are allowed to have a discrete center; for example, SL(2, **R**) has a center of order 2, but is still counted as a simple Lie group. If the center is non-trivial (and not the whole group) then the simple Lie group is not simple as an abstract group. Some authors require that the center of a simple Lie group be finite (or trivial); the universal cover of SL(2, **R**) is an example of a simple Lie group with infinite center.

- **R**: Usually the group **R** of real numbers under addition (and its quotient **R/Z**) are not counted as simple Lie groups, even though they are connected and have a Lie algebra with no proper non-zero ideals. Occasionally authors define simple Lie groups in such a way that **R** is simple, though this sometimes seems to be an accident caused by overlooking this case.

49

- Matrix groups: Some authors restrict themselves to Lie groups that can be represented as groups of finite matrices. The metaplectic group is an example of a simple Lie group that cannot be represented in this way.

- Complex Lie algebras: The definition of a simple Lie algebra is not stable under the *extension of scalars*. The complexification of a complex simple Lie algebra, such as **sl**(n, **C**) is semisimple, but not simple.

The most common definition is the one above: simple Lie groups have to be connected, they are allowed to have non-trivial centers (possibly infinite), they need not be representable by finite matrices, and they must be non-abelian.

5.2   Method of classification

Main article: list of simple Lie groups

Such groups are classified using the prior classification of the complex simple Lie algebras: for which see the page on root systems. It is shown that a simple Lie group has a simple Lie algebra that will occur on the list given there, once it is complexified (that is, made into a complex vector space rather than a real one). This reduces the classification to two further matters.

5.3   Real forms

The groups SO(p,q,**R**) and SO($p+q$,**R**), for example, give rise to different real Lie algebras, but having the same Dynkin diagram. In general there may be different *real forms* of the same complex Lie algebra.

5.4   Relationship of simple Lie algebras to groups

Secondly the Lie algebra only determines uniquely the simply connected (universal) cover G^* of the component containing the identity of a Lie group G. It may well happen that G^* isn't actually a simple group, for example having a non-trivial center. We have therefore to worry about the global topology, by computing the fundamental group of G (an abelian group: a Lie group is an H-space). This was done by Élie Cartan.

For an example, take the special orthogonal groups in even dimension. With the non-identity matrix $-I$ in the center, these aren't actually simple groups; and having a twofold spin cover, they aren't simply-connected either. They lie 'between' G^* and G, in the notation above.

5.5   Classification by Dynkin diagram

Main article: root system

According to Dynkin's classification, we have as possibilities these only, where n is the number of nodes:

5.6 Infinite series

5.6.1 A series

A_1, A_2, ...

A_r corresponds to the special unitary group, SU(r + 1).

5.6.2 B series

B_2, B_3, ...

B_r corresponds to the special orthogonal group, SO($2r$ + 1).

5.6.3 C series

C_3, C_4, ...

C_r corresponds to the symplectic group, Sp(2r).

5.6.4 D series

D_4, D_5, ...

D_r corresponds to the special orthogonal group, SO($2r$), starting with SO(8). The diagram D_2 is two isolated nodes, the same as $A_1 \cup A_1$, and this coincidence corresponds to the covering map homomorphism from SU(2) × SU(2) to SO(4) given by quaternion multiplication; see quaternions and spatial rotation. Thus SO(4) is not a simple group. Also, the diagram D_3 is the same as A_3, corresponding to a covering map homomorphism from SU(4) to SO(6). With D_4 there is an 'exotic' symmetry of the diagram, corresponding to so-called triality.

5.7 Exceptional cases

For the so-called exceptional cases see G_2, F_4, E_6, E_7, and E_8. These cases are deemed 'exceptional' because they do not fall into infinite series of groups of increasing dimension. From the point of view of each group taken separately, there is nothing so unusual about them. These exceptional groups were discovered around 1890 in the classification of the simple Lie algebras, over the complex numbers (Wilhelm Killing, re-done by Élie Cartan). For some time it was a research issue to find concrete ways in which they arise, for example as a symmetry group of a differential system.

See also $E_7\frac{1}{2}$.

5.8 Simply laced groups

A **simply laced group** is a Lie group whose Dynkin diagram only contain simple links, and therefore all the nonzero roots of the corresponding Lie algebra have the same length. The A, D and E series groups are all simply laced, but no group of type B, C, F, or G is simply laced.

5.9 See also

- Cartan matrix
- Coxeter matrix
- Weyl group
- Coxeter group
- Kac–Moody algebra
- Catastrophe theory

5.10 References

- Jacobson, Nathan (1971-06-01). *Exceptional Lie Algebras* (1 ed.). CRC Press. ISBN 0-8247-1326-5.

- Fulton, WIlliam and Harris, Joe. *Representation Theory, A First Course*, Springer Readings in Mathematics

Chapter 6

Table of Lie groups

This article gives a table of some common Lie groups and their associated Lie algebras.

The following are noted: the topological properties of the group (dimension; connectedness; compactness; the nature of the fundamental group; and whether or not they are simply connected) as well as on their algebraic properties (abelian; simple; semisimple).

For more examples of Lie groups and other related topics see the list of simple Lie groups; the Bianchi classification of groups of up to three dimensions; and the list of Lie group topics.

6.1 Real Lie groups and their algebras

Column legend

- **Cpt**: Is this group G compact? (Yes or No)

- π_0 : Gives the group of components of G. The order of the component group gives the number of connected components. The group is connected if and only if the component group is trivial (denoted by 0).

- π_1 : Gives the fundamental group of G whenever G is connected. The group is simply connected if and only if the fundamental group is trivial (denoted by 0).

- **UC**: If G is not simply connected, gives the universal cover of G.

6.2 Real Lie algebras

Table legend:

- **S**: Is this algebra simple? (Yes or No)

- **SS**: Is this algebra semi-simple? (Yes or No)

6.3 Complex Lie groups and their algebras

The dimensions given are dimensions over **C**. Note that every complex Lie group/algebra can also be viewed as a real Lie group/algebra of twice the dimension.

6.4   Complex Lie algebras

The dimensions given are dimensions over **C**. Note that every complex Lie algebra can also be viewed as a real Lie algebra of twice the dimension.

6.5   References

- Fulton, William; Harris, Joe (1991), *Representation theory. A first course*, Graduate Texts in Mathematics, Readings in Mathematics **129**, New York: Springer-Verlag, ISBN 978-0-387-97495-8, MR 1153249, ISBN 978-0-387-97527-6

Chapter 7

Lie algebra

"Lie bracket" redirects here. For the operation on vector fields, see Lie bracket of vector fields.

In mathematics, a **Lie algebra** (/ˈliː/, not /ˈlaɪ/) is a vector space together with a non-associative multiplication called "Lie bracket" $[x, y]$. It was introduced to study the concept of infinitesimal transformations. Hermann Weyl introduced the term "Lie algebra" (after Sophus Lie) in the 1930s. In older texts, the name "**infinitesimal group**" is used.

Lie algebras are closely related to Lie groups which are groups that are also smooth manifolds, with the property that the group operations of multiplication and inversion are smooth maps. Any Lie group gives rise to a Lie algebra. Conversely, to any finite-dimensional Lie algebra over real or complex numbers, there is a corresponding connected Lie group unique up to covering (Lie's third theorem). This correspondence between Lie groups and Lie algebras allows one to study Lie groups in terms of Lie algebras.

7.1 Definitions

A **Lie algebra** is a vector space $\mathfrak{g}$ over some field F together with a binary operation $[\cdot, \cdot] : \mathfrak{g} \times \mathfrak{g} \to \mathfrak{g}$ called the **Lie bracket** that satisfies the following axioms:

- Bilinearity,

$$[ax + by, z] = a[x, z] + b[y, z], \quad [z, ax + by] = a[z, x] + b[z, y]$$

for all scalars a, b in F and all elements x, y, z in $\mathfrak{g}$.

- Alternativity,

$$[x, x] = 0$$

for all x in $\mathfrak{g}$.

- The Jacobi identity,

$$[x, [y, z]] + [z, [x, y]] + [y, [z, x]] = 0$$

for all x, y, z in $\mathfrak{g}$.

Using bilinearity to expand the Lie bracket $[x+y, x+y]$ and using alternativity shows that $[x, y] + [y, x] = 0$ for all elements x, y in $\mathfrak{g}$, showing that bilinearity and alternativity together imply

- Anticommutativity,

    $$[x, y] = -[y, x],$$

    for all elements x, y in $\mathfrak{g}$. Anticommutativity only implies the alternating property if the field's characteristic is not 2.[1]

It is customary to express a Lie algebra in lower-case fraktur, like $\mathfrak{g}$. If a Lie algebra is associated with a Lie group, then the spelling of the Lie algebra is the same as that Lie group. For example, the Lie algebra of SU(n) is written as $\mathfrak{su}(n)$.

7.1.1   Generators and dimension

Elements of a Lie algebra $\mathfrak{g}$ are said to be **generators** of the Lie algebra if the smallest subalgebra of $\mathfrak{g}$ containing them is $\mathfrak{g}$ itself. The **dimension** of a Lie algebra is its dimension as a vector space over F. The cardinality of a minimal generating set of a Lie algebra is always less than or equal to its dimension.

7.1.2   Subalgebras, ideals and homomorphisms

The Lie bracket is not associative in general, meaning that $[[x, y], z]$ need not equal $[x, [y, z]]$. Nonetheless, much of the terminology that was developed in the theory of associative rings or associative algebras is commonly applied to Lie algebras. A subspace $\mathfrak{h} \subseteq \mathfrak{g}$ that is closed under the Lie bracket is called a **Lie subalgebra**. If a subspace $I \subseteq \mathfrak{g}$ satisfies a stronger condition that

$$[\mathfrak{g}, I] \subseteq I,$$

then I is called an **ideal** in the Lie algebra $\mathfrak{g}$.[2] A **homomorphism** between two Lie algebras (over the same base field) is a linear map that is compatible with the respective Lie brackets:

$$f : \mathfrak{g} \to \mathfrak{g}', \quad f([x, y]) = [f(x), f(y)],$$

for all elements x and y in $\mathfrak{g}$. As in the theory of associative rings, ideals are precisely the kernels of homomorphisms, given a Lie algebra $\mathfrak{g}$ and an ideal I in it, one constructs the **factor algebra** $\mathfrak{g}/I$, and the first isomorphism theorem holds for Lie algebras.

Let S be a subset of $\mathfrak{g}$. The set of elements x such that $[x, s] = 0$ for all s in S forms a subalgebra called the centralizer of S. The centralizer of $\mathfrak{g}$ itself is called the center of $\mathfrak{g}$. Similar to centralizers, if S is a subspace,[3] then the set of x such that $[x, s]$ is in S for all s in S forms a subalgebra called the normalizer of S.

7.1.3   Direct sum and semidirect product

Given two Lie algebras $\mathfrak{g}$ and $\mathfrak{g}'$, their direct sum is the Lie algebra consisting of the vector space $\mathfrak{g} \oplus \mathfrak{g}'$, of the pairs (x, x'), $x \in \mathfrak{g}, x' \in \mathfrak{g}'$, with the operation

$$[(x, x'), (y, y')] = ([x, y], [x', y']), \quad x, y \in \mathfrak{g}, \ x', y' \in \mathfrak{g}'.$$

Let $\mathfrak{g}$ be a Lie algebra and $\mathfrak{i}$ its ideal. If the canonical map $\mathfrak{g} \to \mathfrak{g}/\mathfrak{i}$ splits (i.e., admits a section), then $\mathfrak{g}$ is said to be a semidirect product of $\mathfrak{i}$ and $\mathfrak{g}/\mathfrak{i}$.

Levi's theorem says that a finite-dimensional Lie algebra is a semidirect product of its radical and the complementary subalgebra (Levi subalgebra).

7.2 Properties

7.2.1 Admits an enveloping algebra

See also: Universal enveloping algebra

For any associative algebra A with multiplication $*$, one can construct a Lie algebra $L(A)$. As a vector space, $L(A)$ is the same as A. The Lie bracket of two elements of $L(A)$ is defined to be their commutator in A:

$$[a, b] = a * b - b * a.$$

The associativity of the multiplication $*$ in A implies the Jacobi identity of the commutator in $L(A)$. For example, the associative algebra of $n \times n$ matrices over a field F gives rise to the general linear Lie algebra $\mathfrak{gl}_n(F)$. The associative algebra A is called an **enveloping algebra** of the Lie algebra $L(A)$. Every Lie algebra can be embedded into one that arises from an associative algebra in this fashion; see universal enveloping algebra.

7.2.2 Representation

Given a vector space V, let $\mathfrak{gl}(V)$ denote the Lie algebra enveloped by the associative algebra of all linear endomorphisms of V. A representation of a Lie algebra $\mathfrak{g}$ on V is a Lie algebra homomorphism

$$\pi : \mathfrak{g} \to \mathfrak{gl}(V).$$

A representation is said to be faithful if its kernel is trivial. Every finite-dimensional Lie algebra has a faithful representation on a finite-dimensional vector space (Ado's theorem).[4]

For example,

$$\mathrm{ad} : \mathfrak{g} \to \mathfrak{gl}(\mathfrak{g})$$

given by $\mathrm{ad}(x)(y) = [x, y]$ is a representation of $\mathfrak{g}$ on the vector space $\mathfrak{g}$ called the adjoint representation. A derivation on the Lie algebra $\mathfrak{g}$ (in fact on any non-associative algebra) is a linear map $\delta : \mathfrak{g} \to \mathfrak{g}$ that obeys the Leibniz' law, that is,

$$\delta([x, y]) = [\delta(x), y] + [x, \delta(y)]$$

for all x and y in the algebra. For any x, $\mathrm{ad}(x)$ is a derivation; a consequence of the Jacobi identity. Thus, the image of ad lies in the subalgebra of $\mathfrak{gl}(\mathfrak{g})$ consisting of derivations on $\mathfrak{g}$. A derivation that happens to be in the image of ad is called an inner derivation. If $\mathfrak{g}$ is semisimple, every derivation on $\mathfrak{g}$ is inner.

7.3 Examples

7.3.1 Vector spaces

- Any vector space V endowed with the identically zero Lie bracket becomes a Lie algebra. Such Lie algebras are called abelian, cf. below. Any one-dimensional Lie algebra over a field is abelian, by the antisymmetry of the Lie bracket.

- The real vector space of all $n \times n$ skew-hermitian matrices is closed under the commutator and forms a real Lie algebra denoted $\mathfrak{u}(n)$. This is the Lie algebra of the unitary group $U(n)$.

7.3.2   Subspaces

- The subspace of the general linear Lie algebra $\mathfrak{gl}_n(F)$ consisting of matrices of trace zero is a subalgebra,[5] the special linear Lie algebra, denoted $\mathfrak{sl}_n(F)$.

7.3.3   Real matrix groups

- Any Lie group G defines an associated real Lie algebra $\mathfrak{g}$ =Lie(G). The definition in general is somewhat technical, but in the case of real matrix groups, it can be formulated via the exponential map, or the matrix exponent. The Lie algebra $\mathfrak{g}$ consists of those matrices X for which $\exp(tX) \in G$, $\forall$ real numbers t.

  The Lie bracket of $\mathfrak{g}$ is given by the commutator of matrices. As a concrete example, consider the special linear group SL(n,**R**), consisting of all $n \times n$ matrices with real entries and determinant 1. This is a matrix Lie group, and its Lie algebra consists of all $n \times n$ matrices with real entries and trace 0.

7.3.4   Three dimensions

- The three-dimensional Euclidean space $\mathbf{R}^3$ with the Lie bracket given by the cross product of vectors becomes a three-dimensional Lie algebra.

- The Heisenberg algebra $H_3(\mathrm{R})$ is a three-dimensional Lie algebra generated by elements x, y and z with Lie brackets

$$[x, y] = z, \quad [x, z] = 0, \quad [y, z] = 0$$

It is explicitly realized as the space of 3×3 strictly upper-triangular matrices, with the Lie bracket given by the matrix commutator,

$$x = \begin{pmatrix} 0 & 1 & 0 \\ 0 & 0 & 0 \\ 0 & 0 & 0 \end{pmatrix}, \quad y = \begin{pmatrix} 0 & 0 & 0 \\ 0 & 0 & 1 \\ 0 & 0 & 0 \end{pmatrix}, \quad z = \begin{pmatrix} 0 & 0 & 1 \\ 0 & 0 & 0 \\ 0 & 0 & 0 \end{pmatrix}.$$

Any element of the Heisenberg group is thus representable as a product of group generators, i.e., matrix exponentials of these Lie algebra generators,

$$\begin{pmatrix} 1 & a & c \\ 0 & 1 & b \\ 0 & 0 & 1 \end{pmatrix} = e^{by} e^{cz} e^{ax}.$$

- The commutation relations between the x, y, and z components of the angular momentum operator in quantum mechanics are the same as those of $\mathfrak{su}(2)$ and $\mathfrak{so}(3)$,

$$[L_x, L_y] = i\hbar L_z$$

$$[L_y, L_z] = i\hbar L_x$$

$$[L_z, L_x] = i\hbar L_y$$

(The physicist convention for Lie algebras is used in the above equations, hence the factor of i.) The Lie algebra formed by these operators have, in fact, representations of all finite dimensions.

7.3.5 Infinite dimensions

- An important class of infinite-dimensional real Lie algebras arises in differential topology. The space of smooth vector fields on a differentiable manifold M forms a Lie algebra, where the Lie bracket is defined to be the commutator of vector fields. One way of expressing the Lie bracket is through the formalism of Lie derivatives, which identifies a vector field X with a first order partial differential operator LX acting on smooth functions by letting $LX(f)$ be the directional derivative of the function f in the direction of X. The Lie bracket $[X,Y]$ of two vector fields is the vector field defined through its action on functions by the formula:

$$L_{[X,Y]}f = L_X(L_Y f) - L_Y(L_X f).$$

- A Kac–Moody algebra is an example of an infinite-dimensional Lie algebra.

- The Moyal algebra is an infinite-dimensional Lie algebra which contains all classical Lie algebras as subalgebras.

7.4 Structure theory and classification

Lie algebras can be classified to some extent. In particular, this has an application to the classification of Lie groups.

7.4.1 Abelian, nilpotent, and solvable

Analogously to abelian, nilpotent, and solvable groups, defined in terms of the derived subgroups, one can define abelian, nilpotent, and solvable Lie algebras.

A Lie algebra $\mathfrak{g}$ is **abelian** if the Lie bracket vanishes, i.e. $[x,y] = 0$, for all x and y in $\mathfrak{g}$. Abelian Lie algebras correspond to commutative (or abelian) connected Lie groups such as vector spaces K^n or tori T^n, and are all of the form $\mathfrak{k}^n$, meaning an n-dimensional vector space with the trivial Lie bracket.

A more general class of Lie algebras is defined by the vanishing of all commutators of given length. A Lie algebra $\mathfrak{g}$ is **nilpotent** if the lower central series

$$\mathfrak{g} > [\mathfrak{g},\mathfrak{g}] > [[\mathfrak{g},\mathfrak{g}],\mathfrak{g}] > [[[\mathfrak{g},\mathfrak{g}],\mathfrak{g}],\mathfrak{g}] > \cdots$$

becomes zero eventually. By Engel's theorem, a Lie algebra is nilpotent if and only if for every u in $\mathfrak{g}$ the adjoint endomorphism

$$\mathrm{ad}(u) : \mathfrak{g} \to \mathfrak{g}, \quad \mathrm{ad}(u)v = [u,v]$$

is nilpotent.

More generally still, a Lie algebra $\mathfrak{g}$ is said to be **solvable** if the derived series:

$$\mathfrak{g} > [\mathfrak{g},\mathfrak{g}] > [[\mathfrak{g},\mathfrak{g}],[\mathfrak{g},\mathfrak{g}]] > [[[\mathfrak{g},\mathfrak{g}],[\mathfrak{g},\mathfrak{g}]],[[\mathfrak{g},\mathfrak{g}],[\mathfrak{g},\mathfrak{g}]]] > \cdots$$

becomes zero eventually.

Every finite-dimensional Lie algebra has a unique maximal solvable ideal, called its radical. Under the Lie correspondence, nilpotent (respectively, solvable) connected Lie groups correspond to nilpotent (respectively, solvable) Lie algebras.

7.4.2   Simple and semisimple

A Lie algebra is "simple" if it has no non-trivial ideals and is not abelian. A Lie algebra $\mathfrak{g}$ is called **semisimple** if its radical is zero. Equivalently, $\mathfrak{g}$ is semisimple if it does not contain any non-zero abelian ideals. In particular, a simple Lie algebra is semisimple. Conversely, it can be proven that any semisimple Lie algebra is the direct sum of its minimal ideals, which are canonically determined simple Lie algebras.

The concept of semisimplicity for Lie algebras is closely related with the complete reducibility (semisimplicity) of their representations. When the ground field F has characteristic zero, any finite-dimensional representation of a semisimple Lie algebra is semisimple (i.e., direct sum of irreducible representations.) In general, a Lie algebra is called reductive if the adjoint representation is semisimple. Thus, a semisimple Lie algebra is reductive.

7.4.3   Cartan's criterion

Cartan's criterion gives conditions for a Lie algebra to be nilpotent, solvable, or semisimple. It is based on the notion of the Killing form, a symmetric bilinear form on $\mathfrak{g}$ defined by the formula

$$K(u, v) = \mathrm{tr}(\mathrm{ad}(u)\,\mathrm{ad}(v)),$$

where tr denotes the trace of a linear operator. A Lie algebra $\mathfrak{g}$ is semisimple if and only if the Killing form is nondegenerate. A Lie algebra $\mathfrak{g}$ is solvable if and only if $K(\mathfrak{g}, [\mathfrak{g}, \mathfrak{g}]) = 0$.

7.4.4   Classification

The Levi decomposition expresses an arbitrary Lie algebra as a semidirect sum of its solvable radical and a semisimple Lie algebra, almost in a canonical way. Furthermore, semisimple Lie algebras over an algebraically closed field have been completely classified through their root systems. However, the classification of solvable Lie algebras is a 'wild' problem, and cannot be accomplished in general.

7.5   Relation to Lie groups

See also: Lie group–Lie algebra correspondence

Although Lie algebras are often studied in their own right, historically they arose as a means to study Lie groups.

Lie's fundamental theorems describe a relation between Lie groups and Lie algebras. In particular, any Lie group gives rise to a canonically determined Lie algebra (concretely, *the tangent space at the identity*); and, conversely, for any finite-dimensional Lie algebra there is a corresponding connected Lie group (Lie's third theorem; see the Baker–Campbell–Hausdorff formula). This Lie group is not determined uniquely; however, any two connected Lie groups with the same Lie algebra are *locally isomorphic*, and in particular, have the same universal cover. For instance, the special orthogonal group SO(3) and the special unitary group SU(2) give rise to the same Lie algebra, which is isomorphic to $\mathbf{R}^3$ with the cross-product, while SU(2) is a simply-connected twofold cover of SO(3).

Given a Lie group, a Lie algebra can be associated to it either by endowing the tangent space to the identity with the differential of the adjoint map, or by considering the left-invariant vector fields as mentioned in the examples. In the case of real matrix groups, the Lie algebra $\mathfrak{g}$ consists of those matrices X for which $\exp(tX) \in G$ for all real numbers t, where exp is the exponential map.

Some examples of Lie algebras corresponding to Lie groups are the following:

- The Lie algebra $\mathfrak{gl}_n(\mathbb{C})$ for the group $\mathrm{GL}_n(\mathbb{C})$ is the algebra of complex $n{\times}n$ matrices

- The Lie algebra $\mathfrak{sl}_n(\mathbb{C})$ for the group $SL_n(\mathbb{C})$ is the algebra of complex $n \times n$ matrices with trace 0

- The Lie algebras $\mathfrak{o}_n(\mathbb{R})$ for the group $O_n(\mathbb{R})$ and $\mathfrak{so}_n(\mathbb{R})$ for $SO_n(\mathbb{R})$ are both the algebra of real anti-symmetric $n \times n$ matrices (See Antisymmetric matrix: Infinitesimal rotations for a discussion)

- The Lie algebra $\mathfrak{u}_n(\mathbb{C})$ for the group $U_n(\mathbb{C})$ is the algebra of skew-Hermitian complex $n \times n$ matrices while the Lie algebra $\mathfrak{su}_n(\mathbb{C})$ for $SU_n(\mathbb{C})$ is the algebra of skew-Hermitian, traceless complex $n \times n$ matrices.

In the above examples, the Lie bracket $[X, Y]$ (for X and Y matrices in the Lie algebra) is defined as $[X, Y] = XY - YX$
.

Given a set of generators T^a, the **structure constants** f^{abc} express the Lie brackets of pairs of generators as linear combinations of generators from the set, i.e., $[T^a, T^b] = f^{abc} T^c$. The structure constants determine the Lie brackets of elements of the Lie algebra, and consequently nearly completely determine the group structure of the Lie group. The structure of the Lie group near the identity element is displayed explicitly by the Baker–Campbell–Hausdorff formula, an expansion in Lie algebra elements X, Y and their Lie brackets, all nested together within a single exponent, $\exp(tX)$ $\exp(tY) = \exp(tX + tY + \frac{1}{2} t^2 [X, Y] + O(t^3))$).

The mapping from Lie groups to Lie algebras is functorial, which implies that homomorphisms of Lie groups lift to homomorphisms of Lie algebras, and various properties are satisfied by this lifting: it commutes with composition, it maps Lie subgroups, kernels, quotients and cokernels of Lie groups to subalgebras, kernels, quotients and cokernels of Lie algebras, respectively.

The functor **L** which takes each Lie group to its Lie algebra and each homomorphism to its differential is faithful and exact. It is however not an equivalence of categories: different Lie groups may have isomorphic Lie algebras (for example SO(3) and SU(2)), and there are (infinite dimensional) Lie algebras that are not associated to any Lie group.[6]

However, when the Lie algebra $\mathfrak{g}$ is finite-dimensional, one can associate to it a simply connected Lie group having $\mathfrak{g}$ as its Lie algebra. More precisely, the Lie algebra functor **L** has a left adjoint functor $\mathbf{\Gamma}$ from finite-dimensional (real) Lie algebras to Lie groups, factoring through the full subcategory of simply connected Lie groups.[7] In other words, there is a natural isomorphism of bifunctors

$$\mathrm{Hom}(\Gamma(\mathfrak{g}), H) \cong \mathrm{Hom}(\mathfrak{g}, \mathrm{L}(H)).$$

The adjunction $\mathfrak{g} \to \mathrm{L}(\Gamma(\mathfrak{g}))$ (corresponding to the identity on $\Gamma(\mathfrak{g})$) is an isomorphism, and the other adjunction $\Gamma(\mathrm{L}(H)) \to H$ is the projection homomorphism from the universal cover group of the identity component of H to H. It follows immediately that if G is simply connected, then the Lie algebra functor establishes a bijective correspondence between Lie group homomorphisms $G \to H$ and Lie algebra homomorphisms $\mathrm{L}(G) \to \mathrm{L}(H)$.

The universal cover group above can be constructed as the image of the Lie algebra under the exponential map. More generally, we have that the Lie algebra is homeomorphic to a neighborhood of the identity. But globally, if the Lie group is compact, the exponential will not be injective, and if the Lie group is not connected, simply connected or compact, the exponential map need not be surjective.

If the Lie algebra is infinite-dimensional, the issue is more subtle. In many instances, the exponential map is not even locally a homeomorphism (for example, in $\mathrm{Diff}(\mathbf{S}^1)$, one may find diffeomorphisms arbitrarily close to the identity that are not in the image of exp). Furthermore, some infinite-dimensional Lie algebras are not the Lie algebra of any group.

The correspondence between Lie algebras and Lie groups is used in several ways, including in the classification of Lie groups and the related matter of the representation theory of Lie groups. Every representation of a Lie algebra lifts uniquely to a representation of the corresponding connected, simply connected Lie group, and conversely every representation of any Lie group induces a representation of the group's Lie algebra; the representations are in one to one correspondence. Therefore, knowing the representations of a Lie algebra settles the question of representations of the group.

As for classification, it can be shown that any connected Lie group with a given Lie algebra is isomorphic to the universal cover mod a discrete central subgroup. So classifying Lie groups becomes simply a matter of counting the discrete subgroups of the center, once the classification of Lie algebras is known (solved by Cartan et al. in the semisimple case).

7.6   Category theoretic definition

Using the language of category theory, a **Lie algebra** can be defined as an object A in **Vec**k, the category of vector spaces over a field k of characteristic not 2, together with a morphism $[.,.]: A \otimes A \to A$, where $\otimes$ refers to the monoidal product of **Vec**k, such that

- $[\cdot, \cdot] \circ (\mathrm{id} + \tau_{A,A}) = 0$
- $[\cdot, \cdot] \circ ([\cdot, \cdot] \otimes \mathrm{id}) \circ (\mathrm{id} + \sigma + \sigma^2) = 0$

where $\tau\,(a \otimes b) := b \otimes a$ and σ is the cyclic permutation braiding $(\mathrm{id} \otimes \tau_{A,A}) \circ (\tau_{A,A} \otimes \mathrm{id})$. In diagrammatic form:

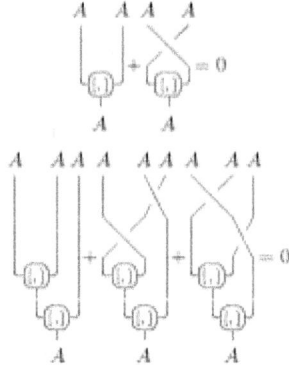

7.7   Lie ring

A **Lie ring** arises as a generalisation of Lie algebras, or through the study of the lower central series of groups. A **Lie ring** is defined as a nonassociative ring with multiplication that is anticommutative and satisfies the Jacobi identity. More specifically we can define a Lie ring L to be an abelian group with an operation $[\cdot, \cdot]$ that has the following properties:

- Bilinearity:

$$[x + y, z] = [x, z] + [y, z], \quad [z, x + y] = [z, x] + [z, y]$$

  for all $x, y, z \in L$.

- The *Jacobi identity*:

$$[x, [y, z]] + [y, [z, x]] + [z, [x, y]] = 0$$

  for all x, y, z in L.

- For all x in L:

$$[x, x] = 0$$

Lie rings need not be Lie groups under addition. Any Lie algebra is an example of a Lie ring. Any associative ring can be made into a Lie ring by defining a bracket operator $[x, y] = xy - yx$. Conversely to any Lie algebra there is a corresponding ring, called the universal enveloping algebra.

Lie rings are used in the study of finite p-groups through the Lazard correspondence. The lower central factors of a p-group are finite abelian p-groups, so modules over $\mathbf{Z}/p\mathbf{Z}$. The direct sum of the lower central factors is given the structure of a Lie ring by defining the bracket to be the commutator of two coset representatives. The Lie ring structure is enriched with another module homomorphism, then pth power map, making the associated Lie ring a so-called restricted Lie ring.

Lie rings are also useful in the definition of a p-adic analytic groups and their endomorphisms by studying Lie algebras over rings of integers such as the p-adic integers. The definition of finite groups of Lie type due to Chevalley involves restricting from a Lie algebra over the complex numbers to a Lie algebra over the integers, and the reducing modulo p to get a Lie algebra over a finite field.

7.7.1 Examples

- Any Lie algebra over a general ring instead of a field is an example of a Lie ring. Lie rings are *not* Lie groups under addition, despite the name.

- Any associative ring can be made into a Lie ring by defining a bracket operator $[x, y] = xy - yx$.

- For an example of a Lie ring arising from the study of groups, let G be a group with $(x, y) = x^{-1}y^{-1}xy$ the commutator operation, and let $G = G_0 \supseteq G_1 \supseteq G_2 \supseteq \cdots \supseteq G_n \supseteq \cdots$ be a central series in G — that is the commutator subgroup (G_i, G_j) is contained in G_{i+j} for any i, j . Then

$$L = \bigoplus G_i / G_{i+1}$$

  is a Lie ring with addition supplied by the group operation (which will be commutative in each homogeneous part), and the bracket operation given by

$$[xG_i, yG_j] = (x, y)G_{i+j}$$

(x, y)

7.8 See also

7.9 Notes

[1] Humphreys p. 1

[2] Due to the anticommutativity of the commutator, the notions of a left and right ideal in a Lie algebra coincide.

[3] Jacobson 1962, pg. 28

[4] Jacobson 1962, Ch. VI

[5] Humphreys p.2

[6] Beltita 2005, pg. 75

[7] Adjoint property is discussed in more general context in Hofman & Morris (2007) (e.g., page 130) but is a straightforward consequence of, e.g., Bourbaki (1989) Theorem 1 of page 305 and Theorem 3 of page 310.

7.10   References

- Beltita, Daniel. *Smooth Homogeneous Structures in Operator Theory*, CRC Press, 2005. ISBN 978-1-4200-3480-6

- Boza, Luis; Fedriani, Eugenio M. & Núñez, Juan. *A new method for classifying complex filiform Lie algebras*, Applied Mathematics and Computation, 121 (2-3): 169–175, 2001

- Bourbaki, Nicolas. "Lie Groups and Lie Algebras - Chapters 1-3", Springer, 1989, ISBN 3-540-64242-0

- Erdmann, Karin & Wildon, Mark. *Introduction to Lie Algebras*, 1st edition, Springer, 2006. ISBN 1-84628-040-0

- Hall, Brian C. *Lie Groups, Lie Algebras, and Representations: An Elementary Introduction*, Springer, 2003. ISBN 0-387-40122-9

- Hofman, Karl & Morris, Sidney. "The Lie Theory of Connected Pro-Lie Groups", European Mathematical Society, 2007, ISBN 978-3-03719-032-6

- Humphreys, James E. *Introduction to Lie Algebras and Representation Theory*, Second printing, revised. Graduate Texts in Mathematics, 9. Springer-Verlag, New York, 1978. ISBN 0-387-90053-5

- Jacobson, Nathan, *Lie algebras*, Republication of the 1962 original. Dover Publications, Inc., New York, 1979. ISBN 0-486-63832-4

- Kac, Victor G. et al. *Course notes for MIT 18.745: Introduction to Lie Algebras*, math.mit.edu

- O'Connor, J.J. & Robertson, E.F. Biography of Sophus Lie, MacTutor History of Mathematics Archive, www-history.mcs.st-andrews.ac.uk

- O'Connor, J.J. & Robertson, E.F. Biography of Wilhelm Killing, MacTutor History of Mathematics Archive, www-history.mcs.st-andrews.ac.uk

- Serre, Jean-Pierre. "Lie Algebras and Lie Groups", 2nd edition, Springer, 2006. ISBN 3-540-55008-9

- Steeb, W.-H. *Continuous Symmetries, Lie Algebras, Differential Equations and Computer Algebra*, second edition, World Scientific, 2007, ISBN 978-981-270-809-0

- Varadarajan, V.S. *Lie Groups, Lie Algebras, and Their Representations*, 1st edition, Springer, 2004. ISBN 0-387-90969-9.

7.11   External links

- Hazewinkel, Michiel, ed. (2001), "Lie algebra", *Encyclopedia of Mathematics*, Springer, ISBN 978-1-55608-010-4

- McKenzie, Douglas, (2015), "An Elementary Introduction to Lie Algebras for Physicists"

Chapter 8

Semisimple Lie algebra

In mathematics, a Lie algebra is **semisimple** if it is a direct sum of simple Lie algebras, i.e., non-abelian Lie algebras $\mathfrak{g}$ whose only ideals are $\{0\}$ and $\mathfrak{g}$ itself.

Throughout the article, unless otherwise stated, $\mathfrak{g}$ is a finite-dimensional Lie algebra over a field of characteristic 0. The following conditions are equivalent:

- $\mathfrak{g}$ is semisimple

- the Killing form, $\kappa(x,y) = \mathrm{tr}(\mathrm{ad}(x)\mathrm{ad}(y))$, is non-degenerate,

- $\mathfrak{g}$ has no non-zero abelian ideals,

- $\mathfrak{g}$ has no non-zero solvable ideals,

- The radical (maximal solvable ideal) of $\mathfrak{g}$ is zero.

8.1 Examples

Examples of semisimple Lie algebras, with notation coming from classification by Dynkin diagrams, are:

- $A_n : \mathfrak{sl}_{n+1}$, the special linear Lie algebra.

- $B_n : \mathfrak{so}_{2n+1}$, the odd-dimensional special orthogonal Lie algebra.

- $C_n : \mathfrak{sp}_{2n}$, the symplectic Lie algebra.

- $D_n : \mathfrak{so}_{2n}$, the even-dimensional special orthogonal Lie algebra.

These Lie algebras are numbered so that n is the rank. Except certain exceptions in low dimensions, many of these are simple Lie algebras, which are *a fortiori* semisimple. These four families, together with five exceptions (E_6, E_7, E_8, F_4, and G_2), are in fact the *only* simple Lie algebras over the complex numbers.

8.2 Classification

See also: Root system

Every semisimple Lie algebra over an algebraically closed field is a direct sum of simple Lie algebras (by definition), and the finite-dimensional simple Lie algebras fall in four families – A_n, B_n, C_n, and D_n – with five exceptions E_6, E_7, E_8, F_4, and G_2. Simple Lie algebras are classified by the connected Dynkin diagrams, shown on the right, while semisimple Lie algebras correspond to not necessarily connected Dynkin diagrams, where each component of the diagram corresponds to a summand of the decomposition of the semisimple Lie algebra into simple Lie algebras.

The classification proceeds by considering a Cartan subalgebra (maximal abelian Lie algebra; corresponds to a maximal torus in a Lie group) and the adjoint action of the Lie algebra on this subalgebra. The root system of the action then both determines the original Lie algebra and must have a very constrained form, which can be classified by the Dynkin diagrams.

The classification is widely considered one of the most elegant results in mathematics – a brief list of axioms yields, via a relatively short proof, a complete but non-trivial classification with surprising structure. This should be compared to the classification of finite simple groups, which is significantly more complicated.

The enumeration of the four families is non-redundant and consists only of simple algebras if $n \geq 1$ for A_n, $n \geq 2$ for B_n, $n \geq 3$ for C_n, and $n \geq 4$ for D_n. If one starts numbering lower, the enumeration is redundant, and one has exceptional isomorphisms between simple Lie algebras, which are reflected in isomorphisms of Dynkin diagrams; the E_n can also be extended down, but below E_6 are isomorphic to other, non-exceptional algebras.

Over a non-algebraically closed field, the classification is more complicated – one classifies simple Lie algebras over the algebraic closure, then for each of these, one classifies simple Lie algebras over the original field which have this form (over the closure). For example, to classify simple real Lie algebras, one classifies real Lie algebras with a given complexification, which are known as real forms of the complex Lie algebra; this can be done by Satake diagrams, which are Dynkin diagrams with additional data ("decorations").

8.3   History

The semisimple Lie algebras over the complex numbers were first classified by Wilhelm Killing (1888–90), though his proof lacked rigor. His proof was made rigorous by Élie Cartan (1894) in his Ph.D. thesis, who also classified semisimple real Lie algebras. This was subsequently refined, and the present classification by Dynkin diagrams was given by then 22-year-old Eugene Dynkin in 1947. Some minor modifications have been made (notably by J. P. Serre), but the proof is unchanged in its essentials and can be found in any standard reference, such as (Humphreys 1972).

8.4   Properties

8.4.1   Complete reducibility

A consequence of semisimplicity is a theorem due to Weyl: every finite-dimensional representation is completely reducible; that is for every invariant subspace of the representation there is an invariant complement. Infinite-dimensional representations of semisimple Lie algebras are not in general completely reducible.

8.4.2   Centerless

Since the center of a Lie algebra $\mathfrak{g}$ is an abelian ideal, if $\mathfrak{g}$ is semisimple, then its center is zero. (Note: since $\mathfrak{gl}_n$ has non-trivial center, it is not semisimple.) In other words, the adjoint representation ad is injective. Moreover, it can be shown that the dimension of the Lie algebra $\mathrm{Der}(\mathfrak{g})$ of derivations on $\mathfrak{g}$ is equal to the dimension of $\mathfrak{g}$. Hence, $\mathfrak{g}$ is Lie algebra isomorphic to $\mathrm{Der}(\mathfrak{g})$. (This is a special case of Whitehead's lemma.) Every ideal, quotient and product of semisimple Lie algebras is again semisimple.

8.4.3   Linear

The adjoint representation is injective, and so a semisimple Lie algebra is also a linear Lie algebra under the adjoint representation. This may lead to some ambiguity, as every Lie algebra is already linear with respect to some other vector space (Ado's theorem), although not necessarily via the adjoint representation. But in practice, such ambiguity rarely occurs.

8.4.4 Jordan decomposition

Any endomorphism x of a finite-dimensional vector space over an algebraically closed field can be decomposed uniquely into a diagonalizable (or semisimple) and nilpotent part

$$x = s + n$$

such that s and n commute with each other. Moreover, each of s and n is a polynomial in x. This is a consequence of the Jordan decomposition.

If $x \in \mathfrak{g}$, then the image of x under the adjoint map decomposes as

$$\mathrm{ad}(x) = \mathrm{ad}(s) + \mathrm{ad}(n).$$

The elements s and n are *unique* elements of $\mathfrak{g}$ such that n is nilpotent, s is semisimple, n and s commute, and for which such a decomposition holds. This abstract Jordan decomposition factors through any representation of $\mathfrak{g}$ in the sense that given any representation ρ,

$$\rho(x) = \rho(s) + \rho(n)$$

is the Jordan decomposition of $\rho(x)$ in the endomorphism ring of the representation space.

8.4.5 Rank

The **rank** of a complex semisimple Lie algebra is the dimension of any of its Cartan subalgebras.

8.5 Significance

The significance of semisimplicity comes firstly from the Levi decomposition, which states that every finite dimensional Lie algebra is the semidirect product of a solvable ideal (its radical) and a semisimple algebra. In particular, there is no nonzero Lie algebra that is both solvable and semisimple.

Semisimple Lie algebras have a very elegant classification, in stark contrast to solvable Lie algebras. Semisimple Lie algebras over an algebraically closed field are completely classified by their root system, which are in turn classified by Dynkin diagrams. Semisimple algebras over non-algebraically closed fields can be understood in terms of those over the algebraic closure, though the classification is somewhat more intricate; see real form for the case of real semisimple Lie algebras, which were classified by Élie Cartan.

Further, the representation theory of semisimple Lie algebras is much cleaner than that for general Lie algebras. For example, the Jordan decomposition in a semisimple Lie algebra coincides with the Jordan decomposition in its representation; this is not the case for Lie algebras in general.

If $\mathfrak{g}$ is semisimple, then $\mathfrak{g} = [\mathfrak{g}, \mathfrak{g}]$. In particular, every linear semisimple Lie algebra is a subalgebra of $\mathfrak{sl}$, the special linear Lie algebra. The study of the structure of $\mathfrak{sl}$ constitutes an important part of the representation theory for semisimple Lie algebras.

8.6   Generalizations

Main articles: Reductive Lie algebra and Split Lie algebra

Semisimple Lie algebras admit certain generalizations. Firstly, many statements that are true for semisimple Lie algebras are true more generally for reductive Lie algebras. Abstractly, a reductive Lie algebra is one whose adjoint representation is completely reducible, while concretely, a reductive Lie algebra is a direct sum of a semisimple Lie algebra and an abelian Lie algebra; for example, $\mathfrak{sl}_n$ is semisimple, and $\mathfrak{gl}_n$ is reductive. Many properties of semisimple Lie algebras depend only on reducibility.

Many properties of complex semisimple/reductive Lie algebras are true not only for semisimple/reductive Lie algebras over algebraically closed fields, but more generally for split semisimple/reductive Lie algebras over other fields: semisimple/reductive Lie algebras over algebraically closed fields are always split, but over other fields this is not always the case. Split Lie algebras have essentially the same representation theory as semsimple Lie algebras over algebraically closed fields, for instance, the splitting Cartan subalgebra playing the same role as the Cartan subalgebra plays over algebraically closed fields. This is the approach followed in (Bourbaki 2005), for instance, which classifies representations of split semisimple/reductive Lie algebras.

8.7   References

- Bourbaki, Nicolas (2005), "VIII: Split Semi-simple Lie Algebras", *Elements of Mathematics: Lie Groups and Lie Algebras: Chapters 7–9*

- Erdmann, Karin; Wildon, Mark (2006), *Introduction to Lie Algebras* (1st ed.), Springer, ISBN 1-84628-040-0.

- Humphreys, James E. (1972), *Introduction to Lie Algebras and Representation Theory*, Berlin, New York: Springer-Verlag, ISBN 978-0-387-90053-7.

- Varadarajan, V. S. (2004), *Lie Groups, Lie Algebras, and Their Representations* (1st ed.), Springer, ISBN 0-387-90969-9.

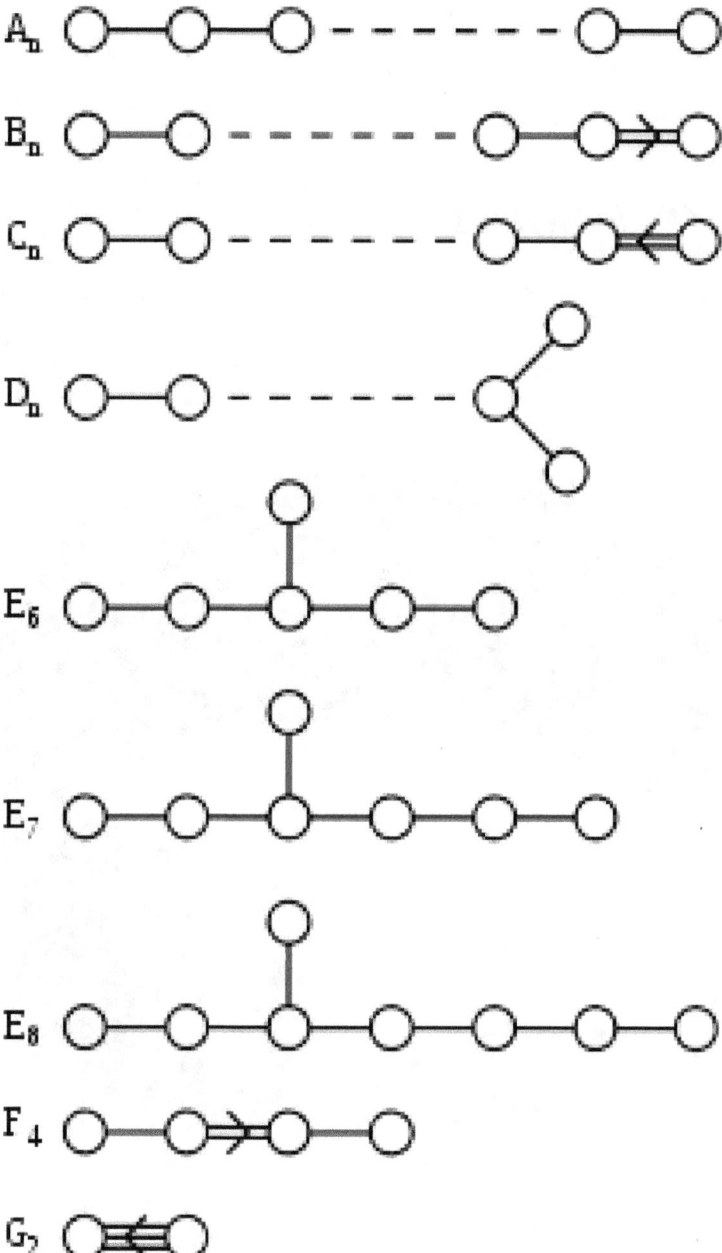

The simple Lie algebras are classified by the connected Dynkin diagrams.

Chapter 9

Homogeneous space

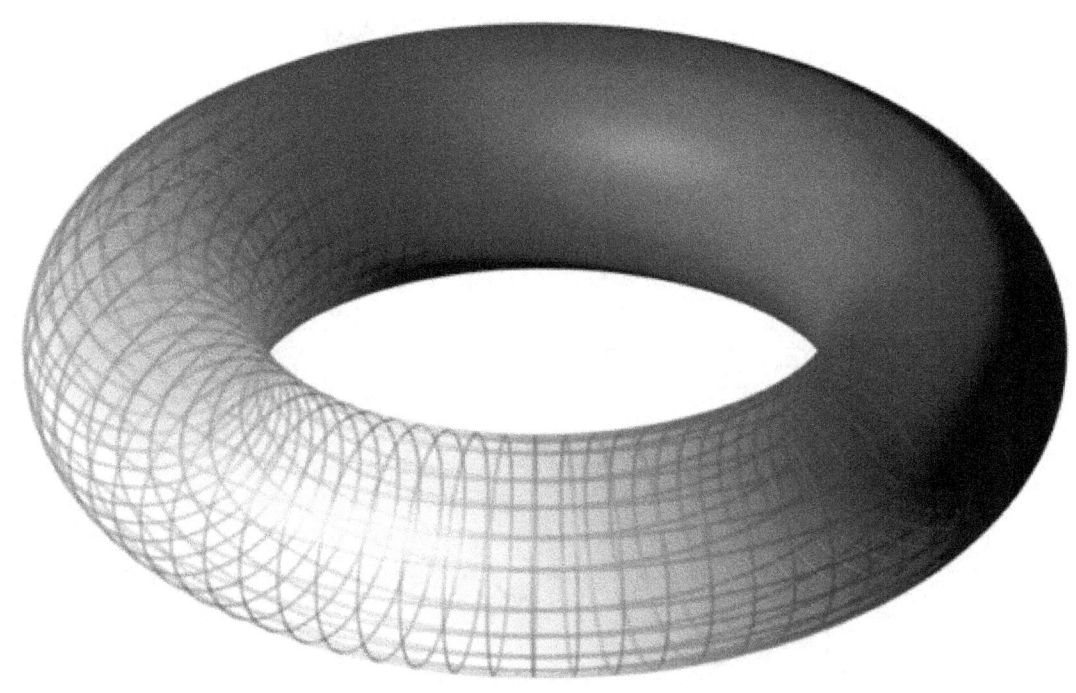

A torus. The standard torus is homogeneous under its diffeomorphism and homeomorphism groups, and the flat torus is homogeneous under its diffeomorphism, homeomorphism, and isometry groups.

In mathematics, particularly in the theories of Lie groups, algebraic groups and topological groups, a **homogeneous space** for a group G is a non-empty manifold or topological space X on which G acts transitively. The elements of G are called the **symmetries** of X. A special case of this is when the group G in question is the automorphism group of the space X – here "automorphism group" can mean isometry group, diffeomorphism group, or homeomorphism group. In this case X is homogeneous if intuitively X looks locally the same at each point, either in the sense of isometry (rigid geometry), diffeomorphism (differential geometry), or homeomorphism (topology). Some authors insist that the action of G be faithful (non-identity elements act non-trivially), although the present article does not. Thus there is a group action of G on X which can be thought of as preserving some "geometric structure" on X, and making X into a single G-orbit.

9.1 Formal definition

Let X be a non-empty set and G a group. Then X is called a G-space if it is equipped with an action of G on X.[1] Note that automatically G acts by automorphisms (bijections) on the set. If X in addition belongs to some category, then the elements of G are assumed to act as automorphisms in the same category. Thus the maps on X effected by G are structure preserving. A homogeneous space is a G-space on which G acts transitively.

Succinctly, if X is an object of the category $\mathbf{C}$, then the structure of a G-space is a homomorphism:

$$\rho : G \to \mathrm{Aut}_{\mathbf{C}}(X)$$

into the group of automorphisms of the object X in the category $\mathbf{C}$. The pair (X, ρ) defines a homogeneous space provided $\rho(G)$ is a transitive group of symmetries of the underlying set of X.

9.1.1 Examples

For example, if X is a topological space, then group elements are assumed to act as homeomorphisms on X. The structure of a G-space is a group homomorphism $\rho : G \to \mathrm{Homeo}(X)$ into the homeomorphism group of X.

Similarly, if X is a differentiable manifold, then the group elements are diffeomorphisms. The structure of a G-space is a group homomorphism $\rho : G \to \mathrm{Diffeo}(X)$ into the diffeomorphism group of X.

Riemannian symmetric spaces are an important class of homogeneous spaces, and include many of the examples listed below.

Concrete examples include:

Isometry groups

- Positive curvature:

1. Sphere (orthogonal group): $S^{n-1} \cong \mathrm{O}(n)/\mathrm{O}(n-1)$

2. Oriented sphere (special orthogonal group): $S^{n-1} \cong \mathrm{SO}(n)/\mathrm{SO}(n-1)$

3. Projective space (projective orthogonal group): $\mathbf{P}^{n-1} \cong \mathrm{PO}(n)/\mathrm{PO}(n-1)$

- Flat (zero curvature):

1. Euclidean space (Euclidean group, point stabilizer is orthogonal group): $\mathbf{A}^n \cong \mathrm{E}(n)/\mathrm{O}(n)$

- Negative curvature:

1. Hyperbolic space (orthochronous Lorentz group, point stabilizer orthogonal group, corresponding to hyperboloid model): $\mathbf{H}^n \cong \mathrm{O}^+(1, n)/\mathrm{O}(n)$

2. Oriented hyperbolic space: $\mathrm{SO}^+(1, n)/\mathrm{SO}(n)$

3. Anti-de Sitter space: $\mathrm{AdS}_{n+1} = \mathrm{O}(2, n)/\mathrm{O}(1, n)$

Others

- Affine space (for affine group, point stabilizer general linear group): $\mathbf{A}^n = \mathrm{Aff}(n, K)/\mathrm{GL}(n, k)$.

- Grassmannian: $\mathrm{Gr}(r, n) = \mathrm{O}(n)/(\mathrm{O}(r) \times \mathrm{O}(n-r))$

9.2   Geometry

From the point of view of the Erlangen program, one may understand that "all points are the same", in the geometry of X. This was true of essentially all geometries proposed before Riemannian geometry, in the middle of the nineteenth century.

Thus, for example, Euclidean space, affine space and projective space are all in natural ways homogeneous spaces for their respective symmetry groups. The same is true of the models found of non-Euclidean geometry of constant curvature, such as hyperbolic space.

A further classical example is the space of lines in projective space of three dimensions (equivalently, the space of two-dimensional subspaces of a four-dimensional vector space). It is simple linear algebra to show that GL_4 acts transitively on those. We can parameterize them by *line co-ordinates*: these are the 2×2 minors of the 4×2 matrix with columns two basis vectors for the subspace. The geometry of the resulting homogeneous space is the line geometry of Julius Plücker.

9.3   Homogeneous spaces as coset spaces

In general, if X is a homogeneous space, and Ho is the stabilizer of some marked point o in X (a choice of origin), the points of X correspond to the left cosets G/Ho, and the marked point o corresponds to the coset of the identity. Conversely, given a coset space G/H, it is a homogeneous space for G with a distinguished point, namely the coset of the identity. Thus a homogeneous space can be thought of as a coset space without a choice of origin.

In general, a different choice of origin o will lead to a quotient of G by a different subgroup Ho' which is related to Ho by an inner automorphism of G. Specifically,

$$H_{o'} = gH_og^{-1} \qquad (1)$$

where g is any element of G for which $go = o'$. Note that the inner automorphism (1) does not depend on which such g is selected; it depends only on g modulo Ho.

If the action of G on X is continuous, then H is a closed subgroup of G. In particular, if G is a Lie group, then H is a Lie subgroup by Cartan's theorem. Hence G/H is a smooth manifold and so X carries a unique smooth structure compatible with the group action.

If H is the identity subgroup $\{e\}$, then X is a principal homogeneous space.

One can go further to *double* coset spaces, notably Clifford–Klein forms $\Gamma\backslash G/H$, where Γ is a discrete subgroup (of G) acting properly discontinuously.

9.4   Example

For example in the line geometry case, we can identify H as a 12-dimensional subgroup of the 16-dimensional general linear group, GL(4), defined by conditions on the matrix entries

$$h_{13} = h_{14} = h_{23} = h_{24} = 0,$$

by looking for the stabilizer of the subspace spanned by the first two standard basis vectors. That shows that X has dimension 4.

Since the homogeneous coordinates given by the minors are 6 in number, this means that the latter are not independent of each other. In fact a single quadratic relation holds between the six minors, as was known to nineteenth-century geometers.

This example was the first known example of a Grassmannian, other than a projective space. There are many further homogeneous spaces of the classical linear groups in common use in mathematics.

9.5 Prehomogeneous vector spaces

The idea of a prehomogeneous vector space was introduced by Mikio Sato.

It is a finite-dimensional vector space V with a group action of an algebraic group G, such that there is an orbit of G that is open for the Zariski topology (and so, dense). An example is GL(1) acting on a one-dimensional space.

The definition is more restrictive than it initially appears: such spaces have remarkable properties, and there is a classification of irreducible prehomogeneous vector spaces, up to a transformation known as "castling".

9.6 Homogeneous spaces in physics

Cosmology using the general theory of relativity makes use of the Bianchi classification system. Homogeneous spaces in relativity represent the space part of background metrics for some cosmological models; for example, the three cases of the Friedmann–Lemaître–Robertson–Walker metric may be represented by subsets of the Bianchi I (flat), V (open), VII (flat or open) and IX (closed) types, while the Mixmaster universe represents an anisotropic example of a Bianchi IX cosmology.[2]

A homogeneous space of N dimensions admits a set of $\frac{1}{2}N(N+1)$ Killing vectors.[3] For three dimensions, this gives a total of six linearly independent Killing vector fields; homogeneous 3-spaces have the property that one may use linear combinations of these to find three everywhere non-vanishing Killing vector fields $\xi_i^{(a)}$,

$$\xi_{[i;k]}^{(a)} = C_{bc}^a \xi_i^{(b)} \xi_k^{(c)}$$

where the object C_{bc}^a, the "structure constants", form a constant order-three tensor antisymmetric in its lower two indices (on the left-hand side, the brackets denote antisymmetrisation and ";" represents the covariant differential operator). In the case of a flat isotropic universe, one possibility is $C_{bc}^a = 0$ (type I), but in the case of a closed FLRW universe, $C_{bc}^a = \varepsilon_{bc}^a$ where ε_{bc}^a is the Levi-Civita symbol.

9.7 See also

- Erlangen program
- Klein geometry
- Heap (mathematics)
- Homogeneous variety

9.8 References

[1] We assume that the action is on the *left*. The distinction is only important in the description of X as a coset space.

[2] Lev Landau and Evgeny Lifshitz (1980), *Course of Theoretical Physics vol. 2: The Classical Theory of Fields*, Butterworth-Heinemann, ISBN 978-0-7506-2768-9

[3] Steven Weinberg (1972), *Gravitation and Cosmology*, John Wiley and Sons

Chapter 10

Representation theory

This article is about the theory of representations of algebraic structures by linear transformations and matrices. For representation theory in other disciplines, see Representation.

Representation theory is a branch of mathematics that studies abstract algebraic structures by *representing* their elements as linear transformations of vector spaces, and studies modules over these abstract algebraic structures.[1] In essence, a representation makes an abstract algebraic object more concrete by describing its elements by matrices and the algebraic operations in terms of matrix addition and matrix multiplication. The algebraic objects amenable to such a description include groups, associative algebras and Lie algebras. The most prominent of these (and historically the first) is the representation theory of groups, in which elements of a group are represented by invertible matrices in such a way that the group operation is matrix multiplication.[2]

Representation theory is a useful method because it reduces problems in abstract algebra to problems in linear algebra, a subject that is well understood.[3] Furthermore, the vector space on which a group (for example) is represented can be infinite-dimensional, and by allowing it to be, for instance, a Hilbert space, methods of analysis can be applied to the theory of groups.[4] Representation theory is also important in physics because, for example, it describes how the symmetry group of a physical system affects the solutions of equations describing that system.[5]

A feature of representation theory is its pervasiveness in mathematics. There are two sides to this. First, the applications of representation theory are diverse:[6] in addition to its impact on algebra, representation theory:

- illuminates and generalizes Fourier analysis via harmonic analysis,[7]

- is connected to geometry via invariant theory and the Erlangen program,[8]

- has an impact in number theory via automorphic forms and the Langlands program.[9]

The second aspect is the diversity of approaches to representation theory. The same objects can be studied using methods from algebraic geometry, module theory, analytic number theory, differential geometry, operator theory, algebraic combinatorics and topology.[10]

The success of representation theory has led to numerous generalizations. One of the most general is in category theory.[11] The algebraic objects to which representation theory applies can be viewed as particular kinds of categories, and the representations as functors from the object category to the category of vector spaces. This description points to two obvious generalizations: first, the algebraic objects can be replaced by more general categories; second, the target category of vector spaces can be replaced by other well-understood categories.

A *representation* should not be confused with a *presentation*.

10.1 Definitions and concepts

Let V be a vector space over a field $\mathbf{F}$.[3] For instance, suppose V is $\mathbf{R}^n$ or $\mathbf{C}^n$, the standard n-dimensional space of column vectors over the real or complex numbers respectively. In this case, the idea of representation theory is to do abstract algebra concretely by using $n \times n$ matrices of real or complex numbers.

There are three main sorts of algebraic objects for which this can be done: groups, associative algebras and Lie algebras.[12]

- The set of all *invertible* $n \times n$ matrices is a group under matrix multiplication and the representation theory of groups analyses a group by describing ("representing") its elements in terms of invertible matrices.

- Matrix addition and multiplication make the set of *all* $n \times n$ matrices into an associative algebra and hence there is a corresponding representation theory of associative algebras.

- If we replace matrix multiplication MN by the matrix commutator $MN - NM$, then the $n \times n$ matrices become instead a Lie algebra, leading to a representation theory of Lie algebras.

This generalizes to any field $\mathbf{F}$ and any vector space V over $\mathbf{F}$, with linear maps replacing matrices and composition replacing matrix multiplication: there is a group $\mathrm{GL}(V,\mathbf{F})$ of automorphisms of V, an associative algebra $\mathrm{End}\mathbf{F}(V)$ of all endomorphisms of V, and a corresponding Lie algebra $\mathbf{gl}(V,\mathbf{F})$.

10.1.1 Definition

See also: group representation, algebra representation and Lie algebra representation

There are two ways to say what a representation is.[13] The first uses the idea of an action, generalizing the way that matrices act on column vectors by matrix multiplication. A representation of a group G or (associative or Lie) algebra A on a vector space V is a map

$$\Phi \colon G \times V \to V \quad \text{or} \quad \Phi \colon A \times V \to V$$

with two properties. First, for any g in G (or a in A), the map

$$\varphi(g) \colon V \to V$$
$$v \mapsto \Phi(g, v)$$

is linear (over $\mathbf{F}$). Second, if we introduce the notation $g \cdot v$ for $\Phi(g, v)$, then for any g_1, g_2 in G and v in V:

(1) $e \cdot v = v$

(2) $g_1 \cdot (g_2 \cdot v) = (g_1 g_2) \cdot v$

where e is the identity element of G and $g_1 g_2$ is product in G. The requirement for associative algebras is analogous, except that associative algebras do not always have an identity element, in which case equation (1) is ignored. Equation (2) is an abstract expression of the associativity of matrix multiplication. This doesn't hold for the matrix commutator and also there is no identity element for the commutator. Hence for Lie algebras, the only requirement is that for any x_1, x_2 in A and v in V:

(2') $x_1 \cdot (x_2 \cdot v) - x_2 \cdot (x_1 \cdot v) = [x_1, x_2] \cdot v$

where $[x_1, x_2]$ is the Lie bracket, which generalizes the matrix commutator $MN - NM$.

The second way to define a representation focuses on the map φ sending g in G to a linear map $\varphi(g)$: $V \to V$, which satisfies

$$\varphi(g_1 g_2) = \varphi(g_1) \circ \varphi(g_2) \quad \text{all for} g_1, g_2 \in G$$

and similarly in the other cases. This approach is both more concise and more abstract. From this point of view:

- a representation of a group G on a vector space V is a group homomorphism φ: $G \to \mathrm{GL}(V,\mathbf{F})$;

- a representation of an associative algebra A on a vector space V is an algebra homomorphism φ: $A \to \mathrm{End}\mathbf{F}(V)$;

- a representation of a Lie algebra $\mathbf{a}$ on a vector space V is a Lie algebra homomorphism φ: $\mathbf{a} \to \mathbf{gl}(V,\mathbf{F})$.

10.1.2   Terminology

The vector space V is called the **representation space** of φ and its dimension (if finite) is called the **dimension** of the representation (sometimes *degree*, as in [14]). It is also common practice to refer to V itself as the representation when the homomorphism φ is clear from the context; otherwise the notation (V,φ) can be used to denote a representation.

When V is of finite dimension n, one can choose a basis for V to identify V with $\mathbf{F}^n$ and hence recover a matrix representation with entries in the field $\mathbf{F}$.

An effective or faithful representation is a representation (V,φ) for which the homomorphism φ is injective.

10.1.3   Equivariant maps and isomorphisms

See also: Equivariant map

If V and W are vector spaces over $\mathbf{F}$, equipped with representations φ and ψ of a group G, then an equivariant map from V to W is a linear map α: $V \to W$ such that

$$\alpha(g \cdot v) = g \cdot \alpha(v)$$

for all g in G and v in V. In terms of φ: $G \to \mathrm{GL}(V)$ and ψ: $G \to \mathrm{GL}(W)$, this means

$$\alpha \circ \phi(g) = \psi(g) \circ \alpha$$

for all g in G.

Equivariant maps for representations of an associative or Lie algebra are defined similarly. If α is invertible, then it is said to be an isomorphism, in which case V and W (or, more precisely, φ and ψ) are *isomorphic representations*.

Isomorphic representations are, for all practical purposes, "the same": they provide the same information about the group or algebra being represented. Representation theory therefore seeks to classify representations "up to isomorphism".

10.1.4   Subrepresentations, quotients, and irreducible representations

See also: Irreducible representation and simple module

If (W,ψ) is a representation of (say) a group G, and V is a linear subspace of W that is preserved by the action of G in the sense that $g \cdot v \in V$ for all $v \in V$ (Serre [14] calls these V *stable under G*), then V is called a *subrepresentation*: by defining

$\varphi(g)$ to be the restriction of $\psi(g)$ to V, (V, φ) is a representation of G and the inclusion of V into W is an equivariant map. The quotient space W/V can also be made into a representation of G.

If W has exactly two subrepresentations, namely the trivial subspace $\{0\}$ and W itself, then the representation is said to be *irreducible*; if W has a proper nontrivial subrepresentation, the representation is said to be *reducible*.[15]

The definition of an irreducible representation implies Schur's lemma: an equivariant map $\alpha\colon V \to W$ between irreducible representations is either the zero map or an isomorphism, since its kernel and image are subrepresentations. In particular, when $V = W$, this shows that the equivariant endomorphisms of V form an associative division algebra over the underlying field $\mathbf{F}$. If $\mathbf{F}$ is algebraically closed, the only equivariant endomorphisms of an irreducible representation are the scalar multiples of the identity.

Irreducible representations are the building blocks of representation theory: if a representation W is not irreducible then it is built from a subrepresentation and a quotient that are both "simpler" in some sense; for instance, if W is finite-dimensional, then both the subrepresentation and the quotient have smaller dimension.

10.1.5 Direct sums and indecomposable representations

See also: Direct sum, indecomposable module and semisimple module

If (V,φ) and (W,ψ) are representations of (say) a group G, then the direct sum of V and W is a representation, in a canonical way, via the equation

$$g \cdot (v, w) = (g \cdot v, g \cdot w).$$

The direct sum of two representations carries no more information about the group G than the two representations do individually. If a representation is the direct sum of two proper nontrivial subrepresentations, it is said to be decomposable. Otherwise, it is said to be indecomposable.

In favourable circumstances, every representation is a direct sum of irreducible representations: such representations are said to be semisimple. In this case, it suffices to understand only the irreducible representations. In other cases, one must understand how indecomposable representations can be built from irreducible representations as extensions of a quotient by a subrepresentation.

10.2 Branches and topics

See also: Group representation

Representation theory is notable for the number of branches it has, and the diversity of the approaches to studying representations of groups and algebras. Although, all the theories have in common the basic concepts discussed already, they differ considerably in detail. The differences are at least 3-fold:

1. Representation theory depends upon the type of algebraic object being represented. There are several different classes of groups, associative algebras and Lie algebras, and their representation theories all have an individual flavour.

2. Representation theory depends upon the nature of the vector space on which the algebraic object is represented. The most important distinction is between finite-dimensional representations and infinite-dimensional ones. In the infinite-dimensional case, additional structures are important (e.g. whether or not the space is a Hilbert space, Banach space, etc.). Additional algebraic structures can also be imposed in the finite-dimensional case.

3. Representation theory depends upon the type of field over which the vector space is defined. The most important case is the field of complex numbers. The other important cases are the field of real numbers, finite fields, and

fields of p-adic numbers. Additional difficulties arise for fields of positive characteristic and for fields that are not algebraically closed.

10.2.1 Finite groups

Main article: Representation of a finite group

Group representations are a very important tool in the study of finite groups.[16] They also arise in the applications of finite group theory to geometry and crystallography.[17] Representations of finite groups exhibit many of the features of the general theory and point the way to other branches and topics in representation theory.

Over a field of characteristic zero, the representation theory of a finite group G has a number of convenient properties. First, the representations of G are semisimple (completely reducible). This is a consequence of Maschke's theorem, which states that any subrepresentation V of a G-representation W has a G-invariant complement. One proof is to choose any projection π from W to V and replace it by its average πG defined by

$$\pi_G(x) = \frac{1}{|G|} \sum_{g \in G} g \cdot \pi(g^{-1} \cdot x).$$

πG is equivariant, and its kernel is the required complement.

The finite-dimensional G-representations can be understood using character theory: the character of a representation φ: $G \to \mathrm{GL}(V)$ is the class function $\chi\varphi$: $G \to \mathbf{F}$ defined by

$$\chi_\varphi(g) = \mathrm{Tr}(\varphi(g))$$

where Tr is the trace. An irreducible representation of G is completely determined by its character.

Maschke's theorem holds more generally for fields of positive characteristic p, such as the finite fields, as long as the prime p is coprime to the order of G. When p and $|G|$ have a common factor, there are G-representations that are not semisimple, which are studied in a subbranch called modular representation theory.

Averaging techniques also show that if $\mathbf{F}$ is the real or complex numbers, then any G-representation preserves an inner product $\langle \cdot, \cdot \rangle$ on V in the sense that

$$\langle g \cdot v, g \cdot w \rangle = \langle v, w \rangle$$

for all g in G and v, w in W. Hence any G-representation is unitary.

Unitary representations are automatically semisimple, since Maschke's result can be proven by taking the orthogonal complement of a subrepresentation. When studying representations of groups that are not finite, the unitary representations provide a good generalization of the real and complex representations of a finite group.

Results such as Maschke's theorem and the unitary property that rely on averaging can be generalized to more general groups by replacing the average with an integral, provided that a suitable notion of integral can be defined. This can be done for compact groups or locally compact groups, using Haar measure, and the resulting theory is known as abstract harmonic analysis.

Over arbitrary fields, another class of finite groups that have a good representation theory are the finite groups of Lie type. Important examples are linear algebraic groups over finite fields. The representation theory of linear algebraic groups and Lie groups extends these examples to infinite-dimensional groups, the latter being intimately related to Lie algebra representations. The importance of character theory for finite groups has an analogue in the theory of weights for representations of Lie groups and Lie algebras.

Representations of a finite group G are also linked directly to algebra representations via the group algebra $\mathbf{F}[G]$, which is a vector space over $\mathbf{F}$ with the elements of G as a basis, equipped with the multiplication operation defined by the group operation, linearity, and the requirement that the group operation and scalar multiplication commute.

10.2.2 Modular representations

Main article: Modular representation theory

Modular representations of a finite group G are representations over a field whose characteristic is not coprime to |G|, so that Maschke's theorem no longer holds (because |G| is not invertible in **F** and so one cannot divide by it).[18] Nevertheless, Richard Brauer extended much of character theory to modular representations, and this theory played an important role in early progress towards the classification of finite simple groups, especially for simple groups whose characterization was not amenable to purely group-theoretic methods because their Sylow 2-subgroups were "too small".[19]

As well as having applications to group theory, modular representations arise naturally in other branches of mathematics, such as algebraic geometry, coding theory, combinatorics and number theory.

10.2.3 Unitary representations

Main article: Unitary representation

A unitary representation of a group G is a linear representation φ of G on a real or (usually) complex Hilbert space V such that $\varphi(g)$ is a unitary operator for every $g \in G$. Such representations have been widely applied in quantum mechanics since the 1920s, thanks in particular to the influence of Hermann Weyl,[20] and this has inspired the development of the theory, most notably through the analysis of representations of the Poincaré group by Eugene Wigner.[21] One of the pioneers in constructing a general theory of unitary representations (for any group G rather than just for particular groups useful in applications) was George Mackey, and an extensive theory was developed by Harish-Chandra and others in the 1950s and 1960s.[22]

A major goal is to describe the "unitary dual", the space of irreducible unitary representations of G.[23] The theory is most well-developed in the case that G is a locally compact (Hausdorff) topological group and the representations are strongly continuous.[7] For G abelian, the unitary dual is just the space of characters, while for G compact, the Peter–Weyl theorem shows that the irreducible unitary representations are finite-dimensional and the unitary dual is discrete.[24] For example, if G is the circle group S^1, then the characters are given by integers, and the unitary dual is **Z**.

For non-compact G, the question of which representations are unitary is a subtle one. Although irreducible unitary representations must be "admissible" (as Harish-Chandra modules) and it is easy to detect which admissible representations have a nondegenerate invariant sesquilinear form, it is hard to determine when this form is positive definite. An effective description of the unitary dual, even for relatively well-behaved groups such as real reductive Lie groups (discussed below), remains an important open problem in representation theory. It has been solved for many particular groups, such as SL(2,**R**) and the Lorentz group.[25]

10.2.4 Harmonic analysis

Main article: Abstract harmonic analysis

The duality between the circle group S^1 and the integers **Z**, or more generally, between a torus T^n and $\mathbf{Z}^n$ is well known in analysis as the theory of Fourier series, and the Fourier transform similarly expresses the fact that the space of characters on a real vector space is the dual vector space. Thus unitary representation theory and harmonic analysis are intimately related, and abstract harmonic analysis exploits this relationship, by developing the analysis of functions on locally compact topological groups and related spaces.[7]

A major goal is to provide a general form of the Fourier transform and the Plancherel theorem. This is done by constructing a measure on the unitary dual and an isomorphism between the regular representation of G on the space $L^2(G)$ of square integrable functions on G and its representation on the space of L^2 functions on the unitary dual. Pontrjagin duality and the Peter–Weyl theorem achieve this for abelian and compact G respectively.[24][26]

Another approach involves considering all unitary representations, not just the irreducible ones. These form a category, and Tannaka–Krein duality provides a way to recover a compact group from its category of unitary representations.

If the group is neither abelian nor compact, no general theory is known with an analogue of the Plancherel theorem or Fourier inversion, although Alexander Grothendieck extended Tannaka–Krein duality to a relationship between linear algebraic groups and tannakian categories.

Harmonic analysis has also been extended from the analysis of functions on a group G to functions on homogeneous spaces for G. The theory is particularly well developed for symmetric spaces and provides a theory of automorphic forms (discussed below).

10.2.5   Lie groups

Main article: Representation of a Lie group

A Lie group is a group that is also a smooth manifold. Many classical groups of matrices over the real or complex numbers are Lie groups.[27] Many of the groups important in physics and chemistry are Lie groups, and their representation theory is crucial to the application of group theory in those fields.[5]

The representation theory of Lie groups can be developed first by considering the compact groups, to which results of compact representation theory apply.[23] This theory can be extended to finite-dimensional representations of semisimple Lie groups using Weyl's unitary trick: each semisimple real Lie group G has a complexification, which is a complex Lie group G^c, and this complex Lie group has a maximal compact subgroup K. The finite-dimensional representations of G closely correspond to those of K.

A general Lie group is a semidirect product of a solvable Lie group and a semisimple Lie group (the Levi decomposition).[28] The classification of representations of solvable Lie groups is intractable in general, but often easy in practical cases. Representations of semidirect products can then be analysed by means of general results called *Mackey theory*, which is a generalization of the methods used in Wigner's classification of representations of the Poincaré group.

10.2.6   Lie algebras

Main article: Lie algebra representation

A Lie algebra over a field **F** is a vector space over **F** equipped with a skew-symmetric bilinear operation called the Lie bracket, which satisfies the Jacobi identity. Lie algebras arise in particular as tangent spaces to Lie groups at the identity element, leading to their interpretation as "infinitesimal symmetries".[28] An important approach to the representation theory of Lie groups is to study the corresponding representation theory of Lie algebras, but representations of Lie algebras also have an intrinsic interest.[29]

Lie algebras, like Lie groups, have a Levi decomposition into semisimple and solvable parts, with the representation theory of solvable Lie algebras being intractable in general. In contrast, the finite-dimensional representations of semisimple Lie algebras are completely understood, after work of Élie Cartan. A representation of a semisimple Lie algebra **g** is analysed by choosing a Cartan subalgebra, which is essentially a generic maximal subalgebra **h** of **g** on which the Lie bracket is zero ("abelian"). The representation of **g** can be decomposed into weight spaces that are eigenspaces for the action of **h** and the infinitesimal analogue of characters. The structure of semisimple Lie algebras then reduces the analysis of representations to easily understood combinatorics of the possible weights that can occur.[28]

Infinite-dimensional Lie algebras

See also: Affine Lie algebra and Kac–Moody algebra

There are many classes of infinite-dimensional Lie algebras whose representations have been studied. Among these, an important class are the Kac–Moody algebras.[30] They are named after Victor Kac and Robert Moody, who independently discovered them. These algebras form a generalization of finite-dimensional semisimple Lie algebras, and share many of their combinatorial properties. This means that they have a class of representations that can be understood in the same way as representations of semisimple Lie algebras.

Affine Lie algebras are a special case of Kac–Moody algebras, which have particular importance in mathematics and theoretical physics, especially conformal field theory and the theory of exactly solvable models. Kac discovered an elegant proof of certain combinatorial identities, Macdonald identities, which is based on the representation theory of affine Kac–Moody algebras.

Lie superalgebras

Main article: Representation of a Lie superalgebra

Lie superalgebras are generalizations of Lie algebras in which the underlying vector space has a Z_2-grading, and skew-symmetry and Jacobi identity properties of the Lie bracket are modified by signs. Their representation theory is similar to the representation theory of Lie algebras.[31]

10.2.7 Linear algebraic groups

See also: Linear algebraic group

Linear algebraic groups (or more generally, affine group schemes) are analogues in algebraic geometry of Lie groups, but over more general fields than just **R** or **C**. In particular, over finite fields, they give rise to finite groups of Lie type. Although linear algebraic groups have a classification that is very similar to that of Lie groups, their representation theory is rather different (and much less well understood) and requires different techniques, since the Zariski topology is relatively weak, and techniques from analysis are no longer available.[32]

10.2.8 Invariant theory

Main article: Invariant theory

Invariant theory studies actions on algebraic varieties from the point of view of their effect on functions, which form representations of the group. Classically, the theory dealt with the question of explicit description of polynomial functions that do not change, or are *invariant*, under the transformations from a given linear group. The modern approach analyses the decomposition of these representations into irreducibles.[33]

Invariant theory of infinite groups is inextricably linked with the development of linear algebra, especially, the theories of quadratic forms and determinants. Another subject with strong mutual influence is projective geometry, where invariant theory can be used to organize the subject, and during the 1960s, new life was breathed into the subject by David Mumford in the form of his geometric invariant theory.[34]

The representation theory of semisimple Lie groups has its roots in invariant theory[27] and the strong links between representation theory and algebraic geometry have many parallels in differential geometry, beginning with Felix Klein's Erlangen program and Élie Cartan's connections, which place groups and symmetry at the heart of geometry.[35] Modern developments link representation theory and invariant theory to areas as diverse as holonomy, differential operators and the theory of several complex variables.

10.2.9 Automorphic forms and number theory

Main article: Automorphic form

Automorphic forms are a generalization of modular forms to more general analytic functions, perhaps of several complex variables, with similar transformation properties.[36] The generalization involves replacing the modular group PSL_2 ($\mathbf{R}$) and a chosen congruence subgroup by a semisimple Lie group G and a discrete subgroup Γ. Just as modular forms can be viewed as differential forms on a quotient of the upper half space $H = PSL_2$ ($\mathbf{R}$)/SO(2), automorphic forms can be viewed as differential forms (or similar objects) on $\Gamma\backslash G/K$, where K is (typically) a maximal compact subgroup of G. Some care is required, however, as the quotient typically has singularities. The quotient of a semisimple Lie group by a compact subgroup is a symmetric space and so the theory of automorphic forms is intimately related to harmonic analysis on symmetric spaces.

Before the development of the general theory, many important special cases were worked out in detail, including the Hilbert modular forms and Siegel modular forms. Important results in the theory include the Selberg trace formula and the realization by Robert Langlands that the Riemann-Roch theorem could be applied to calculate the dimension of the space of automorphic forms. The subsequent notion of "automorphic representation" has proved of great technical value for dealing with the case that G is an algebraic group, treated as an adelic algebraic group. As a result, an entire philosophy, the Langlands program has developed around the relation between representation and number theoretic properties of automorphic forms.[37]

10.2.10   Associative algebras

Main article: Algebra representation

In one sense, associative algebra representations generalize both representations of groups and Lie algebras. A representation of a group induces a representation of a corresponding group ring or group algebra, while representations of a Lie algebra correspond bijectively to representations of its universal enveloping algebra. However, the representation theory of general associative algebras does not have all of the nice properties of the representation theory of groups and Lie algebras.

Module theory

Main article: Module theory

When considering representations of an associative algebra, one can forget the underlying field, and simply regard the associative algebra as a ring, and its representations as modules. This approach is surprisingly fruitful: many results in representation theory can be interpreted as special cases of results about modules over a ring.

Hopf algebras and quantum groups

Main article: Representation theory of Hopf algebras

Hopf algebras provide a way to improve the representation theory of associative algebras, while retaining the representation theory of groups and Lie algebras as special cases. In particular, the tensor product of two representations is a representation, as is the dual vector space.

The Hopf algebras associated to groups have a commutative algebra structure, and so general Hopf algebras are known as quantum groups, although this term is often restricted to certain Hopf algebras arising as deformations of groups or their universal enveloping algebras. The representation theory of quantum groups has added surprising insights to the representation theory of Lie groups and Lie algebras, for instance through the crystal basis of Kashiwara.

10.3   Generalizations

10.3.1 Set-theoretic representations

Main article: Group action

A *set-theoretic representation* (also known as a group action or *permutation representation*) of a group G on a set X is given by a function ρ from G to X^X, the set of functions from X to X, such that for all g_1, g_2 in G and all x in X:

$$\rho(1)[x] = x$$

$$\rho(g_1 g_2)[x] = \rho(g_1)[\rho(g_2)[x]].$$

This condition and the axioms for a group imply that $\rho(g)$ is a bijection (or permutation) for all g in G. Thus we may equivalently define a permutation representation to be a group homomorphism from G to the symmetric group SX of X.

10.3.2 Representations in other categories

See also: Category theory

Every group G can be viewed as a category with a single object; morphisms in this category are just the elements of G. Given an arbitrary category C, a *representation* of G in C is a functor from G to C. Such a functor selects an object X in C and a group homomorphism from G to Aut(X), the automorphism group of X.

In the case where C is **VectF**, the category of vector spaces over a field **F**, this definition is equivalent to a linear representation. Likewise, a set-theoretic representation is just a representation of G in the category of sets.

For another example consider the category of topological spaces, **Top**. Representations in **Top** are homomorphisms from G to the homeomorphism group of a topological space X.

Two types of representations closely related to linear representations are:

- projective representations: in the category of projective spaces. These can be described as "linear representations up to scalar transformations".

- affine representations: in the category of affine spaces. For example, the Euclidean group acts affinely upon Euclidean space.

10.3.3 Representations of categories

See also: Quiver (mathematics)

Since groups are categories, one can also consider representation of other categories. The simplest generalization is to monoids, which are categories with one object. Groups are monoids for which every morphism is invertible. General monoids have representations in any category. In the category of sets, these are monoid actions, but monoid representations on vector spaces and other objects can be studied.

More generally, one can relax the assumption that the category being represented has only one object. In full generality, this is simply the theory of functors between categories, and little can be said.

One special case has had a significant impact on representation theory, namely the representation theory of quivers.[11] A quiver is simply a directed graph (with loops and multiple arrows allowed), but it can be made into a category (and also an algebra) by considering paths in the graph. Representations of such categories/algebras have illuminated several aspects of representation theory, for instance by allowing non-semisimple representation theory questions about a group to be reduced in some cases to semisimple representation theory questions about a quiver.

10.4   See also

- Singlet representation

- Doublet representation

- Philosophy of cusp forms

- Representation (mathematics)

- Representation theorem

- List of representation theory topics

- List of harmonic analysis topics

- Galois representation

10.5   Notes

[1] Classic texts on representation theory include Curtis & Reiner (1962) and Serre (1977). Other excellent sources are Fulton & Harris (1991) and Goodman & Wallach (1998).

[2] For the history of the representation theory of finite groups, see Lam (1998). For algebraic and Lie groups, see Borel (2001).

[3] There are many textbooks on vector spaces and linear algebra. For an advanced treatment, see Kostrikin & Manin (1997).

[4] Sally & Vogan 1989.

[5] Sternberg 1994.

[6] Lam 1998, p. 372.

[7] Folland 1995.

[8] Goodman & Wallach 1998, Olver 1999, Sharpe 1997.

[9] Borel & Casselman 1979, Gelbert 1984.

[10] See the previous footnotes and also Borel (2001).

[11] Simson, Skowronski & Assem 2007.

[12] Fulton & Harris 1991, Simson, Skowronski & Assem 2007, Humphreys 1972.

[13] This material can be found in standard textbooks, such as Curtis & Reiner (1962), Fulton & Harris (1991), Goodman & Wallach (1998), Gordon & Liebeck (1993), Humphreys (1972), Jantzen (2003), Knapp (2001) and Serre (1977).

[14] Serre 1977.

[15] The representation {0} of dimension zero is considered to be neither reducible nor irreducible, just like the number 1 is considered to be neither composite nor prime.

[16] Alperin 1986, Lam 1998, Serre 1977.

[17] Kim 1999.

[18] Serre 1977, Part III.

[19] Alperin 1986.

[20] See Weyl 1928.

[21] Wigner 1939.

[22] Borel 2001.

[23] Knapp 2001.

[24] Peter & Weyl 1927.

[25] Bargmann 1947.

[26] Pontrjagin 1934.

[27] Weyl 1946.

[28] Fulton & Harris 1991.

[29] Humphreys 1972a.

[30] Kac 1990.

[31] Kac 1977.

[32] Humphreys 1972b, Jantzen 2003.

[33] Olver 1999.

[34] Mumford, Fogarty & Kirwan 1994.

[35] Sharpe 1997.

[36] Borel & Casselman 1979.

[37] Gelbart 1984.

10.6 References

- Alperin, J. L. (1986), *Local Representation Theory: Modular Representations as an Introduction to the Local Representation Theory of Finite Groups*, Cambridge University Press, ISBN 978-0-521-44926-7.

- Bargmann, V. (1947), "Irreducible unitary representations of the Lorenz group", *Annals of Mathematics* **48** (3): 568–640, doi:10.2307/1969129, JSTOR 1969129.

- Borel, Armand (2001), *Essays in the History of Lie Groups and Algebraic Groups*, American Mathematical Society, ISBN 978-0-8218-0288-5.

- Borel, Armand; Casselman, W. (1979), *Automorphic Forms, Representations, and L-functions*, American Mathematical Society, ISBN 978-0-8218-1435-2.

- Curtis, Charles W.; Reiner, Irving (1962), *Representation Theory of Finite Groups and Associative Algebras*, John Wiley & Sons (Reedition 2006 by AMS Bookstore), ISBN 978-0-470-18975-7.

- Gelbart, Stephen (1984), "An Elementary Introduction to the Langlands Program", *Bulletin of the American Mathematical Society* **10** (2): 177–219, doi:10.1090/S0273-0979-1984-15237-6.

- Folland, Gerald B. (1995), *A Course in Abstract Harmonic Analysis*, CRC Press, ISBN 978-0-8493-8490-5.

- Fulton, William; Harris, Joe (1991), *Representation theory. A first course*, Graduate Texts in Mathematics, Readings in Mathematics **129**, New York: Springer-Verlag, ISBN 978-0-387-97495-8, MR 1153249, ISBN 978-0-387-97527-6.

- Goodman, Roe; Wallach, Nolan R. (1998), *Representations and Invariants of the Classical Groups*, Cambridge University Press, ISBN 978-0-521-66348-9.

- Gordon, James; Liebeck, Martin (1993), *Representations and Characters of Finite Groups*, Cambridge: Cambridge University Press, ISBN 978-0-521-44590-0.

- Helgason, Sigurdur (1978), *Differential Geometry, Lie groups and Symmetric Spaces*, Academic Press, ISBN 978-0-12-338460-7

- Humphreys, James E. (1972a), *Introduction to Lie Algebras and Representation Theory*, Birkhäuser, ISBN 978-0-387-90053-7.

- Humphreys, James E. (1972b), *Linear Algebraic Groups*, Graduate Texts in Mathematics **21**, Berlin, New York: Springer-Verlag, ISBN 978-0-387-90108-4, MR 0396773

- Jantzen, Jens Carsten (2003), *Representations of Algebraic Groups*, American Mathematical Society, ISBN 978-0-8218-3527-2.

- Kac, Victor G. (1977), "Lie superalgebras", *Advances in Mathematics* **26** (1): 8–96, doi:10.1016/0001-8708(77)90017-2.

- Kac, Victor G. (1990), *Infinite Dimensional Lie Algebras* (3rd ed.), Cambridge University Press, ISBN 978-0-521-46693-6.

- Knapp, Anthony W. (2001), *Representation Theory of Semisimple Groups: An Overview Based on Examples*, Princeton University Press, ISBN 978-0-691-09089-4.

- Kim, Shoon Kyung (1999), *Group Theoretical Methods and Applications to Molecules and Crystals: And Applications to Molecules and Crystals*, Cambridge University Press, ISBN 978-0-521-64062-6.

- Kostrikin, A. I.; Manin, Yuri I. (1997), *Linear Algebra and Geometry*, Taylor & Francis, ISBN 978-90-5699-049-7.

- Lam, T. Y. (1998), "Representations of finite groups: a hundred years", *Notices of the AMS* (American Mathematical Society) **45** (3,4): 361–372 (Part I), 465–474 (Part II).

- Yurii I. Lyubich. *Introduction to the Theory of Banach Representations of Groups*. Translated from the 1985 Russian-language edition (Kharkov, Ukraine). Birkhäuser Verlag. 1988.

- Mumford, David; Fogarty, J.; Kirwan, F. (1994), *Geometric invariant theory*, Ergebnisse der Mathematik und ihrer Grenzgebiete (2) [Results in Mathematics and Related Areas (2)] **34** (3rd ed.), Berlin, New York: Springer-Verlag, ISBN 978-3-540-56963-3, MR 0214602(1st ed. 1965); MR 0719371 (2nd ed.); MR 1304906(3rd ed.)

- Olver, Peter J. (1999), *Classical invariant theory*, Cambridge: Cambridge University Press, ISBN 0-521-55821-2.

- Peter, F.; Weyl, Hermann (1927), "Die Vollständigkeit der primitiven Darstellungen einer geschlossenen kontinuierlichen Gruppe", *Mathematische Annalen* **97** (1): 737–755, doi:10.1007/BF01447892.

- Pontrjagin, Lev S. (1934), "The theory of topological commutative groups", *Annals of Mathematics* (Annals of Mathematics) **35** (2): 361–388, doi:10.2307/1968438, JSTOR 1968438.

- Sally, Paul; Vogan, David A. (1989), *Representation Theory and Harmonic Analysis on Semisimple Lie Groups*, American Mathematical Society, ISBN 978-0-8218-1526-7.

- Serre, Jean-Pierre (1977), *Linear Representations of Finite Groups*, Springer-Verlag, ISBN 978-0387901909.

- Sharpe, Richard W. (1997), *Differential Geometry: Cartan's Generalization of Klein's Erlangen Program*, Springer, ISBN 978-0-387-94732-7.

- Simson, Daniel; Skowronski, Andrzej; Assem, Ibrahim (2007), *Elements of the Representation Theory of Associative Algebras*, Cambridge University Press, ISBN 978-0-521-88218-7.

- Sternberg, Shlomo (1994), *Group Theory and Physics*, Cambridge University Press, ISBN 978-0-521-55885-3.

- Weyl, Hermann (1928), *Gruppentheorie und Quantenmechanik* (The Theory of Groups and Quantum Mechanics, translated H.P. Robertson, 1931 ed.), S. Hirzel, Leipzig (reprinted 1950, Dover), ISBN 978-0-486-60269-1.

- Weyl, Hermann (1946), *The Classical Groups: Their Invariants and Representations* (2nd ed.), Princeton University Press (reprinted 1997), ISBN 978-0-691-05756-9.

- Wigner, Eugene P. (1939), "On unitary representations of the inhomogeneous Lorentz group", *Annals of Mathematics* (Annals of Mathematics) **40** (1): 149–204, doi:10.2307/1968551, JSTOR 1968551.

10.7 External links

- Hazewinkel, Michiel, ed. (2001), "Representation theory", *Encyclopedia of Mathematics*, Springer, ISBN 978-1-55608-010-4

Chapter 11

Representation theory of the Lorentz group

The Lorentz group, a Lie group on which special relativity is based, has a wide variety of representations. Many of these representations, both finite-dimensional and infinite-dimensional, are important in theoretical physics in the description of particles in relativistic quantum mechanics, as well as of both particles and quantum fields in quantum field theory.

This representation theory also provides the theoretical ground for the concept of spin, which, for a particle, can be either integer or half-integer in the unit of the reduced Planck constant $\hbar$. Quantum mechanical wave functions representing particles with half-integer spin are called spinors. The classical electromagnetic field has spin as well. It transforms under a representation with spin one.[1] It enters into general relativity because in small enough regions of spacetime, physics is that of special relativity.

The group may also be represented in terms of a set of functions defined on the Riemann sphere. These are the Riemann P-functions, which are expressible as hypergeometric functions. The identity component $SO(3; 1)^+$ of the Lorentz group is isomorphic to the Möbius group, and hence any representation of the Lorentz group is necessarily a representation of the Möbius group and vice versa.The subgroup $SO(3)$ with its representation theory form a simpler theory, but the two are related and both are prominent in theoretical physics as descriptions of spin, angular momentum, and other phenomena related to rotation.

The adopted Lie algebra basis and conventions used are presented here.

11.1   Finite-dimensional representations

Representation theory of groups in general, and Lie groups in particular, is a very rich subject. The full Lorentz group is no exception. The Lorentz group has some properties that makes it "agreeable" and others that make it "not very agreeable" within the context of representation theory. The group is semisimple, and also simple, but is not connected, and none of its components are simply connected. Perhaps most importantly, the Lorentz group is not compact.[2]

For finite-dimensional representations, the presence of semisimplicity means that the Lorentz group can be dealt with the same way as other semisimple groups using a well-developed theory. In addition, all representations are built from the irreducible ones.[3] But, the non-compactness of the Lorentz group, in combination with lack of simple connectedness, cannot be dealt with in all the aspects as in the simple framework that applies to simply connected, compact groups. Non-compactness implies that no nontrivial finite-dimensional unitary representations exist.[4] Lack of simple connectedness gives rise to spin representations of the group.[5] The non-connectedness means that, for representations of the full Lorentz group, one has to deal with time reversal and space inversion separately.[1][6]

11.1.1   History

The development of the finite-dimensional representation theory of the Lorentz group mostly follows that of the subject in general. Lie theory originated with Sophus Lie in 1873.[7] By 1888 the classification of simple Lie algebras was essentially

Hendrik Antoon Lorentz (right) after whom the Lorentz group is named and Albert Einstein whose special theory of relativity is the main source of application. Photo taken by Paul Ehrenfest 1921.

completed by Wilhelm Killing.[8] In 1913 the theorem of highest weight for representations of simple Lie algebras, the path that will be followed here, was completed by Élie Cartan.[9] Richard Brauer was 1935–38 largely responsible for the development of the Weyl-Brauer matrices describing how spin representations of the Lorentz Lie algebra can be embedded in Clifford algebras.[10] The Lorentz group has also historically received special attention in representation theory, see History of infinite-dimensional unitary representations below, due to its exceptional importance in physics. Hermann Weyl,[11][12] Harish-Chandra[13] mathematicians who also made major contributions to the general theory, and the physicists Eugene Wigner[14] and Valentine Bargmann[15][16] made substantial contributions specialized to the Lorentz group over the years.[17] Physicist Paul Dirac was perhaps the first to manifestly knit everything together in a practical application of major lasting importance with the Dirac equation in 1928.[18]

11.1.2   The Lie algebra

According to the general representation theory of Lie groups, one first looks for the representations of the complexification, **so**(3; 1)**C** of the Lie algebra **so**(3; 1) of the Lorentz group. A convenient basis for **so**(3; 1) is given by the three generators J_i of rotations and the three generators K_i of boosts as described in Conventions below. Now complexify the Lie algebra, and then change basis to the components of:

$$\mathbf{A} = \frac{\mathbf{J}+i\mathbf{K}}{2}, \quad \mathbf{B} = \frac{\mathbf{J}-i\mathbf{K}}{2}. \text{ [19]}$$

In this new basis, one checks that the components of $\mathbf{A} = (A_1, A_2, A_3)$ and $\mathbf{B} = (B_1, B_2, B_3)$ separately satisfy the commutation relations of the Lie algebra **su**(2) and moreover that they commute with each other[20]

$$[A_i, A_j] = i\varepsilon_{ijk}A_k, \quad [B_i, B_j] = i\varepsilon_{ijk}B_k, \quad [A_i, B_j] = 0, \text{ [21]}$$

where i, j, k are indices which each take values 1, 2, 3, and ε_{ijk} is the three-dimensional Levi-Civita symbol. Let **A**C and **B**C denote the complex linear span of **A** and **B** respectively. One has the isomorphisms

$$\mathfrak{so}(3;1) \hookrightarrow \mathfrak{so}(3;1)_C \approx \mathbf{A}_C \oplus \mathbf{B}_C \approx \mathfrak{sl}(2,C) \oplus \mathfrak{sl}(2,C) \approx \mathfrak{sl}(2,C) \oplus i\mathfrak{sl}(2,C) = \mathfrak{sl}(2,C)_C \hookleftarrow \mathfrak{sl}(2,C),$$

where sl(2, **C**) is the complexification of **su**(2) $\approx$ **A** $\approx$ **B**. The utility of these isomorphisms comes from the fact that all irreducible representations of **su**(2) are known. Every irreducible representation of **su**(2) is isomorphic to one of the highest weight representations. Moreover, there is a one-to-one correspondence between linear representations of **su**(2) and complex linear representations of **sl**(2, **C**).[24]

The unitarian trick

In **(A1)**, all isomorphisms are **C**-linear (the last is just a defining equality). The most important part of the manipulations below is that the **R**-linear (irreducible) representations of a (real or complex) Lie algebra are in one-to-one correspondence with **C**-linear (irreducible) representation of its complexification.[25] With this in mind, it is seen that the **R**-linear representations of the **real forms** of the far left, **so**(3; 1), and the far right, **sl**(2, **C**), in **(A1)** can be obtained from the **C**-linear representations of **sl**(2, **C**) $\oplus$ **sl**(2, **C**).

The manipulations to obtain representations of a non-compact algebra (here **so**(3; 1)), and subsequently the non-compact group itself, from qualitative knowledge about unitary representations of a compact group (here SU(2)) is a variant of Weyl's so-called unitarian trick. The trick specialized to SL(2, **C**) can be summarized concisely.[26] Let V be a finite-dimensional complex vector space. The following are equivalent:

- There is a representation of SL(2, **R**) on V

- There is a representation of SU(2) on V

- There is a holomorphic representation of SL(2, **C**) on V

Wilhelm Killing, Independent discoverer of Lie algebras. The simple Lie algebras were first classified by him in 1888.

- There is a representation of $\mathbf{sl}(2, \mathbf{R})$ on V

- There is a representation of $\mathbf{su}(2)$ on V

- There is a complex linear representation of $\mathbf{sl}(2, \mathbf{C})$ on V

Hermann Weyl, the man behind the unitarian trick.

If one representation is irreducible, then all of them are. In this list, cross products (groups) or direct sums (Lie algebras) may be introduced (consistently). The *essence* of the trick is that the starting point in the above list is immaterial. Both qualitative knowledge (like existence theorems for one item on the list) and concrete realizations for one item on the list will translate and propagate, respectively, to the others.

Now, the representations of $\mathbf{sl}(2, \mathbf{C}) \oplus \mathbf{sl}(2, \mathbf{C})$, which is the Lie algebra of $SL(2, \mathbf{C}) \times SL(2, \mathbf{C})$, are supposed to be irreducible. This means that they must be tensor products of complex linear representations of $\mathbf{sl}(2, \mathbf{C})$, as can be seen by restriction to the subgroup $SU(2) \times SU(2) \subset SL(2, \mathbf{C}) \times SL(2, \mathbf{C})$, a compact group to which the Peter–Weyl theorem applies.[27] The irreducible unitary representations of $SU(2) \times SU(2)$ are precisely the tensor products of irreducible unitary representations of $SU(2)$. These stand in one-to-one correspondence with the holomorphic representations of $SL(2, \mathbf{C}) \times SL(2, \mathbf{C})$[27] and these, in turn, are in one-to-one correspondence with the complex linear representations of $\mathbf{sl}(2, \mathbf{C}) \oplus \mathbf{sl}(2, \mathbf{C})$ because $SL(2, \mathbf{C}) \times SL(2, \mathbf{C})$ is simply connected.[27]

For **sl**(2, **C**), there exists the highest weight representations (obtainable, via the trick, from the corresponding **su**(2)-representations), here indexed by μ for μ = 0, 1, ... The tensor products of two complex linear factors then form the irreducible complex linear representations of **sl**(2, **C**) ⊕ **sl**(2, **C**). For reference, if (π_1, U) and (π_2, V) are representations of a Lie algebra **g**, then their tensor product $(\pi_1 \otimes \pi_2, U \otimes V)$ is given by either of

$$\pi_1 \otimes \pi_2(X) = \pi_1(X) \otimes \mathrm{Id}_V + \mathrm{Id}_U \otimes \pi_2(X), \quad X \in \mathbf{g}, \qquad \pi_1 \otimes \pi_2(X, Y) = \pi_1(X) \otimes \mathrm{Id}_V + \mathrm{Id}_U \otimes \pi_2(Y), \quad (X, Y) \in \mathbf{g} \oplus \mathbf{g},$$

where Id is the identity operator. Here, the latter interpretation is intended. The not necessarily complex linear representations of **sl**(2, **C**) come using another variant of the unitarian trick as is shown in the last Lie algebra isomorphism in **(A1)**.

The representations

The representations for all Lie algebras and groups involved in the unitarian trick can now be obtained. The real linear representations for **sl**(2, **C**) and **so**(3; 1) follow here assuming the complex linear representations of **sl**(2, **C**) are known. Explicit realizations and group representations are given later.

sl(2, **C**)   The complex linear representations of the complexification of **sl**(2, **C**), **sl**(2, **C**)C, obtained via isomorphisms in **(A1)**, stand in one-to-one correspondence with the real linear representations of **sl**(2, **C**).[27] The set of all, *at least real linear*, irreducible representations of **sl**(2, **C**) are thus indexed by a pair (μ, ν). The complex linear ones, corresponding precisely to the complexification of the real linear **su**(2) representations, are of the form (μ, 0), while the conjugate linear ones are the (0, ν).[27] All others are real linear only. The linearity properties follow from the canonical injection, the far right in **(A1)**, of **sl**(2, **C**) into its complexification. Representations on the form (ν, ν) or (μ, ν) ⊕ (ν, μ) are given by *real* matrices (the latter is not irreducible). Explicitly, the real linear (μ, ν)-representations of **sl**(2, **C**) are

$$\phi_{\mu,\nu}(X) = \phi_\mu \otimes \overline{\phi}_\nu(X) = \phi_\mu(X) \otimes \mathrm{Id}_{\nu+1} + \mathrm{Id}_{\mu+1} \otimes \overline{\phi_\nu(X)}, \quad X \in \mathfrak{sl}(2, \mathbb{C})$$

where Φ_μ, μ = 0,1, ... are the complex linear irreducible representations of **sl**(2, **C**) and Φ_ν, ν = 0,1, ... their complex conjugate representations. Now the tensor product is interpreted in the former sense of **(A0)**.

so(3; 1)   Via the displayed isomorphisms in (A1) and knowledge of the complex linear irreducible representations of sl(2, **C**) ⊕ sl(2, **C**), upon solving for **J** and **K**, all irreducible representations of **so**(3; 1)**C**, and, by restriction, those of **so**(3; 1) are known. It's worth noting that the representations of **so**(3; 1) obtained this way are real linear (and not complex or conjugate linear) because the algebra is not closed upon conjugation, but they are still irreducible.[22] Since **so**(3; 1) is semisimple,[22] all its representations, not necessarily irreducible, can be built up as direct sums of the irreducible ones.

Thus the finite dimensional irreducible representations of the Lorentz algebra are classified by an ordered pair of half-integers m = μ/2 and n = ν/2, conventionally written as one of

$$(m, n) \equiv D^{(m,n)} \equiv \pi_{m,n}.$$

The notation $D^{(m,n)}$ is usually reserved for the group representations. Let $\pi_{(m, n)} : \mathbf{so}(3; 1) \to \mathbf{gl}(V)$, where V is a vector space, denote the irreducible representations of **so**(3; 1) according to this classification. These are, up to a similarity transformation, uniquely given by

$$\pi_{m,n}(J_i) = \mathbf{1}_{(2m+1)} \otimes J_i^{(n)} + J_i^{(m)} \otimes \mathbf{1}_{(2n+1)}$$

$$\pi_{m,n}(K_i) = i(\mathbf{1}_{(2m+1)} \otimes J_i^{(n)} - J_i^{(m)} \otimes \mathbf{1}_{(2n+1)}),$$

where the $\mathbf{J}^{(n)} = (J^{(n)}_1, J^{(n)}_2, J^{(n)}_3)$ are the (2n + 1)-dimensional irreducible spin n representations of **so**(3) ≈ **su**(2) and $\mathbf{1}_n$ is the n-dimensional unit matrix. Explicit formulas on component form are given at the end of the article.

	$m=0$	$\frac{1}{2}$	1
$n=0$	scalar	Weyl spinor bispinor	self-dual 2-form 2-form field
$\frac{1}{2}$	Weyl spinor (right-handed)	4-vector	Rarita–Schwinger field
1	anti-self-dual 2-form		traceless symmetric tensor

Purple: (m, n) complex irreps   Black: $(m, n) \oplus (n, m)$
Bold: (m, m)

11.1.3 Common representations

Since for any irrep where $m \neq n$ it is essential to operate over the field of complex numbers, the direct sum of representations (m, n) and (n, m) has a particular relevance to physics, since it permits to use linear operators over real numbers.

- $(0, 0)$ is the Lorentz scalar representation. This representation is carried by relativistic scalar field theories.

- $(1/2, 0)$ is the left-handed Weyl spinor and $(0, 1/2)$ is the right-handed Weyl spinor representation.

- $(1/2, 0) \oplus (0, 1/2)$ is the bispinor representation. (See also Dirac spinor and Weyl spinors and bispinors below.)

- $(1/2, 1/2)$ is the four-vector representation. The four-momentum of a particle (either massless or massive) transforms under this representation.

- $(1, 0)$ is the self-dual 2-form field representation and $(0, 1)$ is the anti-self-dual 2-form field representation.

- $(1, 0) \oplus (0, 1)$ is the representation of a parity-invariant 2-form field (a.k.a. curvature form). The electromagnetic field tensor transforms under this representation.

- $(1, 1/2) \oplus (1/2, 1)$ is the Rarita–Schwinger field representation.

- $(1, 1)$ is the spin 2 representation of a traceless symmetric tensor field.[nb 2]

- $(3/2, 0) \oplus (0, 3/2)$ would be the symmetry of the hypothesized gravitino.[nb 3] It can be obtained from the $(1, 1/2) \oplus (1/2, 1)$-representation.[29]

11.1.4 The group

The approach in this section is based on a theorem that, in turn, is based on the fundamental **Lie correspondence**.[30] The Lie correspondence is, in essence, a dictionary between connected Lie groups and Lie algebras.[31]

The Lie correspondence

The Lie correspondence and some results based on it needed here and below are stated for reference. If G denotes a linear Lie group (i.e. a group representable as a group of matrices) and $\mathbf{g}$ a linear Lie algebra (again, linear means representable as an algebra of matrices),[nb 4] let $\Gamma(\mathbf{g})$ denote the group generated by $\exp(\mathbf{g})$, and let $L(G)$ denote the Lie algebra of G (interpreted as the set of matrices X such that $e^{itX} \in G$ for all $t \in \mathbf{R}$). The Lie correspondence reads in modern language, here specialized to linear Lie groups, as follows:

- There is a one-to-one correspondence between connected linear Lie groups and linear Lie algebras given by $G \leftrightarrow \mathbf{g}$ with $\mathbf{g} = L(G)$ or, equivalently $G = \Gamma(\mathbf{g})$, expressed as $\Gamma(L(G)) = G$, respectively $L(\Gamma(\mathbf{g})) = \mathbf{g}$.[32] (Lie)

The following are some corollaries that will be used in the sequel:

- A connected linear Lie group G is abelian if and only if $\mathbf{g}$ is abelian.[33] (Lie i)

- A connected subgroup H with Lie algebra $\mathbf{h}$ of a connected linear Lie group G is normal if and only if $\mathbf{h} \subset \mathbf{g}$ is an ideal.[33] (Lie ii)

- If G, H are linear Lie groups with Lie algebras $\mathbf{g}, \mathbf{h}$ and $\Pi : G \to H$ is a group homomorphism, then $\pi : \mathbf{g} \to \mathbf{h}$, its pushforward at the identity, is a Lie algebra homomorphism and $\Pi(e^{iX}) = e^{i\pi(X)}$ for every $X \in \mathbf{g}$.[34] (Lie iii)

*Sophus Lie, the originator of Lie theory. The theory of manifolds was not discovered in Lie's time, so he worked locally with subsets of R^n. The structure would today be called a **local group**.*

Lie algebra representations from group representations

Using the above theorem it is always possible to pass from a representation of a Lie group G to a representation of its Lie algebra $\mathbf{g}$. If $\Pi : G \to \mathrm{GL}(V)$ is a group representation for some vector space V, then its pushforward (differential) at the identity, or **Lie map**, $\pi : \mathbf{g} \to \mathrm{End}\ V$ is a Lie algebra representation. It is explicitly computed using

$$\pi(X) = \frac{d}{dt}\Pi(e^{itX})|_{t=0}, \quad X \in \mathbf{g}.$$

This, of course, holds for the Lorentz group in particular, but not all Lie algebra representations arise this way because their corresponding group representations may not exist as proper representations, i.e. they are projective, see below.

Group representations from Lie algebra representations

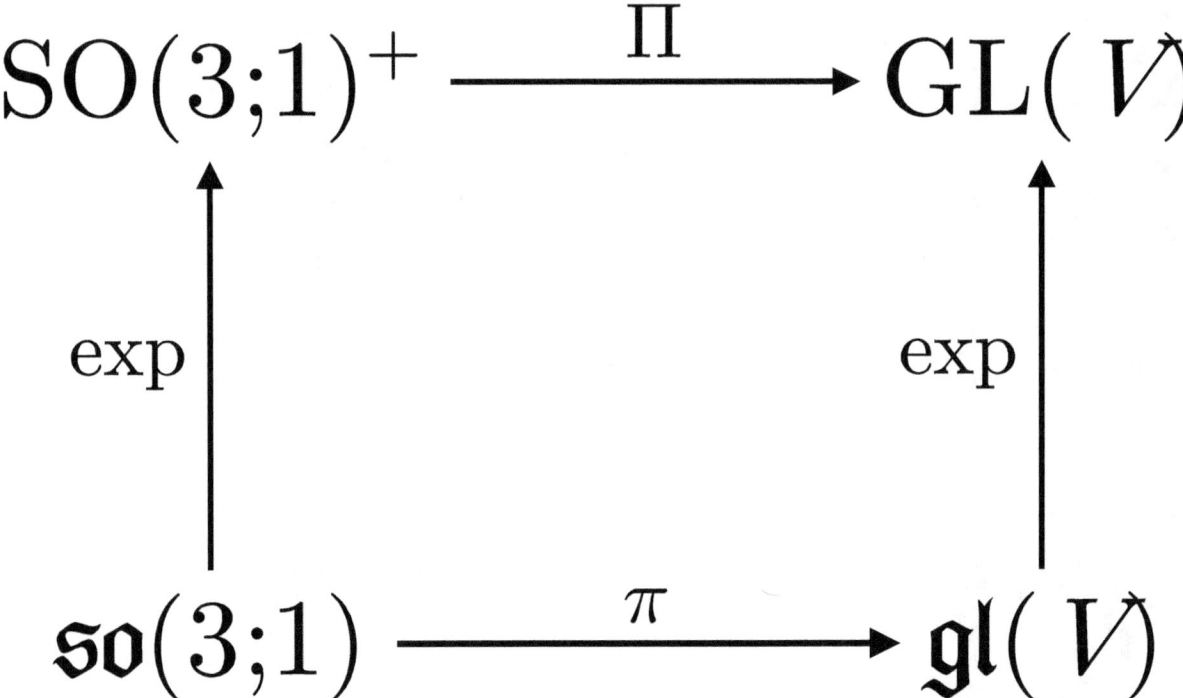

*Here V is a finite-dimensional vector space, GL(V) is the set of all invertible linear transformations on V and **gl**(V) is its Lie algebra. The maps π and Π are Lie algebra and group representations respectively, and exp is the exponential mapping. The diagram commutes only up to a sign if Π is projective.*

Given a **so**(3; 1) representation, one may try to construct a representation of SO(3; 1)$^+$, the identity component of the Lorentz group, by using the exponential mapping. If X is an element of **so**(3; 1) in the standard representation, then

$$\Lambda = e^{iX} \equiv \sum_{n=0}^{\infty} \frac{(iX)^n}{n!}$$

is a Lorentz transformation by general properties of Lie algebras. Motivated by this and the Lie correspondence theorem stated above, let $\pi :$ **so**(3; 1) $\to$ **gl**(V) for some vector space V be a representation and tentatively define a representation Π of SO(3; 1)$^+$ by first setting

$$\Pi_U(e^{iX}) = e^{i\pi(X)}, \quad X \in \mathbf{so}(3;1).$$

The subscript U indicates a small open set containing the identity. Its precise meaning is defined below. There are at least two potential problems with this definition. The first is that it is not obvious that this yields a group homomorphism, or even a well defined map at all (local existence). The second problem is that for a given $g \in U \subset$ SO(3; 1)$^+$ there may not be exactly one $X \in$ **so**(3; 1) such that $g = e^{iX}$ (local uniqueness). The soundness of the tentative definition **(G2)** is given in several steps below:

1. ΠU is a local homomorphism.

2. $\Pi(g)$ defined along a path using properties of ΠU is a global homomorphism.

3. The exponential mapping exp:$\mathbf{so}(3; 1) \to$ SO$(3; 1)^+$ is surjective.

4. $\Pi(g)$ defined along a path coincides with $\Pi U(g)$ with $U =$ SO$(3; 1)^+$.

Local existence and uniqueness   A theorem[35] based on the inverse function theorem states that the map exp : $\mathbf{so}(3;$ 1$) \to$ SO$(3; 1)^+$ is one-to-one for X small enough (**A**). This makes the map well-defined. The qualitative form of the Baker–Campbell–Hausdorff formula then guarantees that it *is* a group homomorphism, still for X small enough (**B**).[36] Let $U \subset$ SO$(3; 1)^+$ denote image under the exponential mapping of the open set in $\mathbf{so}(3; 1)$ where conditions (**A**) and (**B**) both hold. Let $g, h \in U$, $g = e^X$, $h = e^Y$, then

$$\Pi_U(gh) = \Pi_U(e^X e^Y) = e^{i\pi \log(e^X e^Y)} \qquad \text{(by definition and by (A))}$$
$$= e^{i\pi(X+Y+\frac{1}{2}[X,Y]+\frac{1}{12}[X,[X,Y]]-\frac{1}{12}[Y,[X,Y]]+...)} \qquad \text{(by Baker–Campbell–Hausdorff and (B))}$$
$$= e^{i\pi(X)+i\pi(Y)+\frac{1}{2}[i\pi(X),i\pi(Y)]+\frac{1}{12}[i\pi(X),[i\pi(X),i\pi(Y)]]-\frac{1}{12}[i\pi(Y),[i\pi(X),i\pi(Y)]]+...)} \qquad \text{(since π is a Lie algebra homomorphism)}$$
$$= e^{\log(e^{i\pi(X)}e^{i\pi(Y)})} \qquad \text{(by Baker–Campbell–Hausdorff and (B) again)}$$
$$= e^{\log(\Pi_U(g)\Pi_U(h))} = \Pi_U(g)\Pi_U(h) \qquad \text{(by definition and by (A))}$$

This shows that the map ΠU is a well-defined group homomorphism on U.

Global existence and uniqueness   Technically, formula (**G2**) is used to *define* Π near the identity. For other elements $g \notin U$ one chooses a path from the identity to g and defines Π along that path by partitioning it finely enough so that formula (**G2**) can be used again on the resulting factors in the partition. In detail, one sets

$$g = g_n = \left(g_n g_{n-1}^{-1}\right)\left(g_{n-1}g_{n-2}^{-1}\right)\cdots\left(g_2 g_1^{-1}\right)(g_1 g_0), \qquad \Pi(g) \equiv \Pi_U\left(g_n g_{n-1}^{-1}\right)\Pi_U\left(g_{n-1}g_{n-2}^{-1}\right)\cdots\Pi_U\left(g_2 g_1^{-1}\right)\Pi_U(g_1 g_0), \qquad g_0 = 1$$

where the g_i are on the path and the factors on the far right are uniquely defined by (**G2**) provided that all $g_i\, g_{i+1}^{-1} \in U$ and, for all conceivable pairs h,k of points on the path between g_i and g_{i+1}, $hk^{-1} \in U$ as well. For each i take, by the inverse function theorem, the unique X_i such that $\exp(X_i) = g_i g_{i-1}^{-1}$ and obtain

$$\Pi(g) = \Pi_U\left(e^{iX_n}\right)\Pi_U\left(e^{iX_{n-1}}\right)\cdots\Pi_U\left(e^{iX_2}\right)\Pi_U\left(e^{iX_1}\right) = e^{i\pi(X_n)}e^{i\pi(X_{n-1})}\cdots e^{i\pi(X_2)}e^{i\pi(X_1)}.$$

By compactness of the path there is an n large enough so that $\Pi(g)$ is well defined, possibly depending on the partition and/or the path, whether g is close to the identity or not.

Partition independence   It turns out that the result is always independent of the partitioning of the path.[35] To demonstrate the independence of the partitioning of the chosen path, one employs the Baker–Campbell–Hausdorff formula. It shows that ΠU is a group homomorphism for elements in U. To see this, first fix a partitioning used in (**G3**). Then insert a new point h somewhere on the path, say

$$g = \cdots (g_{i+1}h^{-1})(hg_i^{-1})\cdots, \qquad \cdots\Pi_U(g_{i+1}h^{-1})\Pi_U(hg_i^{-1})\cdots.$$

But

$$\cdots\Pi_U(g_{i+1}h^{-1})\Pi_U(hg_i^{-1})\cdots = \cdots\Pi_U(g_{i+1}h^{-1}hg_i^{-1})\cdots = \cdots\Pi_U(g_{i+1}g_i^{-1})\cdots$$

as a consequence of the Baker–Campbell–Hausdorff formula and the conditions on the original partitioning. Thus adding a point on the path has no effect on the definition of $\Pi(g)$. Then for any two given partitionings of a given path, they have common **refinement**, their union. This refinement can be reached from any of the two partitionings by, one-by-one, adding points from the other partition. No individual addition changes the definition of $\Pi(g)$, hence, since there are finitely many points in each partition, the value of $\Pi(g)$ must have been the same for the two partitionings to begin with.

Path independence   For simply connected groups, the construction will be independent of the path as well, yielding a well defined representation.[35] In that case formula **(G2)** can unambiguously be used directly. Simply connected spaces have the property that any two paths can be continuously deformed into each other. Any such deformation is called a homotopy and is usually chosen as a continuous function H from the unit square $\{s,t \in \mathbf{R}: 0 \leq s, t \leq 1\}$ into the group. For $s = 0$ the image is one of the paths, for $s = 1$ the other, for intermediate s, an intermediate path results, but endpoints are kept fixed.

One deforms the path, a little bit at a time, using the previous result, the independence of partitioning. Each consecutive deformation is so small that two consecutive deformed paths can be partitioned using *the same partition points*. Thus two consecutive deformed paths yield the same value for $\Pi(g)$. But any two *pairs* of consecutive deformations need not have the same choice partition points, so the actual path laid out in the group as one progresses through the deformation does indeed change.

Using compactness arguments, in a finite number of steps, the original ($s = 0$) path is deformed into the other ($s = 1$) without affecting the value of $\Pi(g)$.[38]

Global homomorphism   The map ΠU is, by the Baker-Campbell-Hausdorff formula, a local homomorphism. To show that Π is a global homomorphism, consider two elements $g, h \in$ SO(3; 1)$^+$. Lay out paths pg, ph from the identity to them and define a path pgh going along $pg(2t)$ for $0 \leq t \leq 1/2$ and along $pg \cdot ph(2t - 1)$ for $1/2 \leq t \leq 1$. This is a path from the identity to gh. Select adequate partitionings for pg, ph. This corresponds to a choice of "times" $t_0, t_1, ...t_m$ and $s_0, s_1, ...s_n$. Divide the first set with 2 and divide the second set with 2 and add 1/2 and so obtain a new (adequate) set of "times" to be used for pgh. Direct computation shows that, with these partitionings (and hence all partitionings), $\Pi(gh) = \Pi(g)\Pi(h)$.[39]

Surjectiveness of exponential mapping   From a practical point of view, it is important that formula **(G2)** can be used for all elements of the group. The Lie correspondence theorem above guarantees that **(G2)** holds for all $X \in$ **so**(3; 1), but provides no guarantee that all $g \in$ SO(3; 1)$^+$ are in the image of exp:**so**(3; 1) $\rightarrow$ SO(3; 1)$^+$. For general Lie groups, this is not the case, especially not for non-compact groups, as for example for SL(2, **C**), the universal covering group of SO(3; 1)$^+$. It will be treated in this respect below.

But exp: **so**(3; 1) $\rightarrow$ SO(3; 1)$^+$ *is* surjective. One way to see this is to make use of the isomorphism SO(3; 1)$^+ \approx$ PGL(2, **C**), the latter being the Möbius group. It is a quotient of GL(n, **C**) (see the linked article). Let p:GL(n, **C**) $\rightarrow$ PGL(2, **C**) denote the quotient map. Now exp:**gl**(n, **C**) $\rightarrow$ GL(n, **C**) is onto.[40] Now apply the Lie correspondence theorem with π being the differential at the identity of p. Then for all $X \in$ **gl**(n, **C**) $p(e^{iX}) = e^{i\pi(X)}$. Since the left hand side is surjective (both exp and p are), the right hand side is surjective and hence exp:**pgl**(2, **C**) $\rightarrow$ PGL(2, **C**) is surjective. Finally, recycle the argument once more, but now with the known isomorphism between SO(3; 1)$^+$ and PGL(2, **C**) to find that exp is onto for the connected component of the Lorentz group.

Consistency   From the way $\Pi(g)$ has been defined for elements far from the identity, it not immediately clear that formula **(G2)** holds for all elements of SO(3; 1)$^+$, i.e. that one can take $U = G$ in **(G2)**. But, in summary,

- Π is a uniquely constructed homomorphism.

- Using **(G6)** with Π as defined here, then one ends up with the π one started with since Π was defined that way near the identity, and **(G6)** depends only on an arbitrarily small neighborhood of the identity.

- exp: **so**(3; 1) $\rightarrow$ SO(3; 1)$^+$ is surjective.

Hence **(G2)** holds everywhere.[41] One finally unconditionally writes

$$\Pi\left(e^{iX}\right) = e^{i\pi(X)}, \quad X \in \mathfrak{so}(3;1).$$

Projective representations

For a group that is connected but not simply connected, such as SO(3; 1)$^+$, the result *may* depend on the homotopy class of the chosen path.[42] The result, when using **(G2)**, will then depend on *which* X in the Lie algebra is used to obtain the representative matrix for g.

The Lorentz group is doubly connected so that its fundamental group $\pi_1(\text{SO}(3; 1)^+)$, whose elements are the path homotopy classes, has two members. Thus not all representations of the Lie algebra will yield representations of the group, but some will instead yield projective representations.[35] Once these conclusions has been reached, and once one knows whether a representation is projective, there is no need not be concerned about paths and partitions. Formula **(G2)** applies to all group elements and all representations, including the projective ones.

For a projective representation Π of SO(3; 1)$^+$ it holds that

$$[\Pi(\Lambda_1)\Pi(\Lambda_2)\Pi^{-1}(\Lambda_1\Lambda_2)]^2 = 1 \Rightarrow \Pi(\Lambda_1\Lambda_2) = \pm\Pi(\Lambda_1)\Pi(\Lambda_2), \qquad \Lambda_1, \Lambda_2 \in \text{SO}(3;1),$$

since any loop in SO(3; 1)$^+$ traversed twice, due to the double connectedness, is contractible to a point so that its homotopy class is that of a constant map. It follows that Π is a double-valued function. One cannot consistently chose sign to obtain a continuous representation of all of SO(3; 1)$^+$, but this is possible to do *locally* around any point.[4]

11.1.5 The covering group

Consider **sl**(2, **C**) as a *real* Lie algebra with basis $(1/\sqrt{2}\sigma_1, 1/\sqrt{2}\sigma_2, 1/\sqrt{2}\sigma_3, i/\sqrt{2}\sigma_1, i/\sqrt{2}\sigma_2, i/\sqrt{2}\sigma_3) \equiv (j_1, j_2, j_3, k_1, k_2, k_3)$, where the sigmas are the Pauli matrices. From the relations

$$[\sigma_i, \sigma_j] = 2i\epsilon_{ijk}\sigma_k$$

one obtains

$$[j_i, j_j] = i\epsilon_{ijk}j_k, \quad [j_i, k_j] = i\epsilon_{ijk}k_k, \quad [k_i, k_j] = -i\epsilon_{ijk}j_k,$$

which are exactly on the form of the 3-dimensional version of the commutation relations for **so**(3; 1) (see Conventions and Lie algebra bases below). Thus map $Ji \leftrightarrow ji, Ki \leftrightarrow ki$, and extend by linearity to obtain an isomorphism. Since SL(2, **C**) is simply connected, it is the universal covering group of SO(3; 1)$^+$.

A geometric view

Let πg denote the set of path homotopy classes $[pg]$ of paths $pg(t)$, $0 \leq t \leq 1$, from $1 \in \text{SO}(3; 1)^+$ to $g \in \text{SO}(3; 1)^+$ and define the set

$$G = \{(g, [p_g]) : g \in \text{SO}(3;1)^+, [p_g] \in \pi_g\}$$

and endow it with the multiplication operation

$$(g_1, [p_1])(g_2, [p_2]) = (g_1g_2, [p_{12}]), \quad g_1, g_2 \in \text{SO}(3;1)^+, \quad [p_1] \in \pi_{g1}, [p_2] \in \pi_{g2}, [p_{12}] \in \pi_{g12}, \quad p_{12}(t) = p_1(t){\cdot}p_2(t).$$

The dot on the far right denotes path multiplication. With this multiplication, G is a group and $G \approx \text{SL}(2, \mathbf{C})$,[44] the universal covering group of SO(3; 1)$^+$. By the above construction, there is, since each πg has two elements, a 2:1 covering map $p : G \rightarrow \text{SO}(3; 1)^+$ and an isomorphism $G \approx \text{SL}(2, \mathbf{C})$. According to covering group theory, the Lie algebras **so**(3; 1), **sl**(2, **C**) and **g** of G are all isomorphic. The covering map $p{:}G \rightarrow \text{SO}(3; 1)^+$ is simply given by $p(g,[pg]) = g$.

An algebraic view

For an algebraic view of the universal covering group, let SL(2, **C**) act on the set of all Hermitean 2×2 matrices **h** by the operation[45]

$$\mathbf{P}(A): \mathbf{h} \to \mathbf{h}; \quad X \to A^{\dagger} X A, \quad X \in \mathbf{h}, A \in \mathrm{SL}(2, C).$$

Since $X \in \mathbf{h}$ is Hermitean, $A^{\dagger}XA$ is again Hemitean because $(A^{\dagger}XA)^{\dagger} = A^{\dagger}X^{\dagger}A^{\dagger\dagger} = A^{\dagger}XA$, and also $A^{\dagger}(\alpha X + \beta Y)A = \alpha A^{\dagger}XA + \beta A^{\dagger}YA$, so the action is linear as well. An element of **h** may generally be written in the form

$$X = \begin{pmatrix} \xi_4 + \xi_3 & \xi_1 + i\xi_2 \\ \xi_1 - i\xi_2 & \xi_4 - \xi_3 \end{pmatrix}$$

for ξ_i real, showing that **h** is a 4-dimensional real vector space. Moreover, $(AB)^{\dagger}X(AB) = B^{\dagger}A^{\dagger}XAB$ meaning that **P** is a group homomorphism into GL(**h**) ⊂ End **h**. Thus **P** : SL(2, **C**) → GL (**h**) is a 4-dimensional representation of SL(2, **C**). Its kernel must in particular take the identity matrix to itself, $A^{\dagger}IA = A^{\dagger}A = I \Rightarrow A^{\dagger} = A^{-1}$. Thus $AX = XA$ for A in the kernel so, by Schur's lemma,[nb 6] A is a multiple of the identity, which must be $\pm I$ since det $A = 1$.[46] Now map **h** to spacetime $\mathbf{R}^4$ endowed with the Lorentz metric, Minkowski space, via

$$X = (\xi_1, \xi_2, \xi_3, \xi_4) \leftrightarrow \overrightarrow{(\xi_1, \xi_2, \xi_3, \xi_4)} = (x, y, z, t) = \overrightarrow{X}.$$

The action of **P**(A) on **h** preserves determinants since $\det(A^{\dagger}XA) = (\det A)(\det A^{\dagger})(\det X) = \det X$. The induced representation **p** of SL(2, **C**) on $\mathbf{R}^4$, via the above isomorphism, given by

$$\mathbf{p}(A)\overrightarrow{X} = \overrightarrow{AXA^{\dagger}}$$

will preserve the Lorentz inner product since $-\det X = \xi_1^2 + \xi_2^2 + \xi_3^2 - \xi_4^2 = x^2 + y^2 + z^2 - t^2$. This means that **p**(A) belongs to the full Lorentz group SO(3; 1). By the main theorem of connectedness, since SL(2, **C**) is connected, its image under **p** in SO(3; 1) is connected as well, and hence is contained in SO(3; 1)$^+$. It can be shown that the Lie map of **p** : SL(2, **C**) → SO(3; 1)$^+$, π : **sl**(2, **C**) → **so**(3; 1) is a Lie algebra isomorphism (its kernel is $\{\varnothing\}$[nb 7] and must therefore be an isomorphism for dimensional reasons). The map **P** is also onto.[nb 8] Thus SL(2, **C**), since it is simply connected, is the universal covering group of SO(3; 1)$^+$, isomorphic to the group G of above.

Representations of SL(2, C) and sl(2, C)

The complex linear representations of **sl**(2, **C**) and SL(2, **C**) are more straightforward to obtain than the SO(3; 1)$^+$ representations. If $\pi\mu$ is a representation of **su**(2) with highest weight μ, then the complexification of $\pi\mu$ is a complex linear representation of **sl**(2, **C**). All complex linear representation of **sl**(2, **C**) are of this form. The holomorphic group representations (meaning the corresponding Lie algebra representation is complex linear) are obtained by exponentiation. By simple connectedness of SL(2, **C**), this always yields a representation of the group as opposed to in the SO(3; 1)$^+$ case. The real linear representations of **sl**(2, **C**) are exactly the (μ, ν)-representations presented earlier. They can be exponentiated too. The $(\mu, 0)$-representations are complex linear and are (isomorphic to) the highest weight-representations. These are usually indexed with only one integer.

It is also possible to obtain representations of SL(2, **C**) directly. This will be done below. Then, using the unitarian trick, going the other way, one finds **sl**(2, **C**)-,SU(2)-,**su**(2)-,SL(2, **R**)-, and **sl**(2, **R**)-representations as well as **so**(3; 1)-representations (via **(A1)**) and, possibly projective, SO(3; 1)$^+$-representations (via projection from SL(2, **C**), see below, or exponentiation).

The mathematics convention is used in this section for convenience. Lie algebra elements differ by a factor of i and there is no factor of i in the exponential mapping compared to the physics convention used elsewhere. Let the basis of **sl**(2, **C**) be

$$H = \begin{pmatrix} 1 & 0 \\ 0 & -1 \end{pmatrix}, \quad X = \begin{pmatrix} 0 & 1 \\ 0 & 0 \end{pmatrix}, \quad Y = \begin{pmatrix} 0 & 0 \\ 1 & 0 \end{pmatrix}.$$

This choice of basis, and the notation, is standard in the mathematical literature.

Concrete realization   The irreducible holomorphic $(n + 1)$-dimensional representations of SL(2, **C**), $n \geq 0$, can be realized on a set of functions $\mathbb{P}^2 n = \{P\colon \mathbb{C}^2 \to \mathbb{C}\}$ where each $P \in \mathbb{P}^2 n$ is a homogeneous polynomial of degree n in 2 variables.[47][48] The elements of $\mathbb{P}^2 n$ appears as $P(z_1, z_2) = c n z_1{}^n + c_{n-1} z_1{}^{n-1} z_2 + \ldots + c n z_2{}^n$. The action of SL(2, **C**) is given by

$$[\Phi_n(g)]P(z) = \left[\Phi\begin{pmatrix} a & b \\ c & d \end{pmatrix}P\right]\begin{pmatrix} z_1 \\ z_2 \end{pmatrix} = P\left(\begin{pmatrix} a & b \\ c & d \end{pmatrix}^{-1}\begin{pmatrix} z_1 \\ z_2 \end{pmatrix}\right), \quad P \in \mathbb{P}^2_n.$$

The associated **sl**(2, **C**)-action is, using **(G6)** and the definition above, given by

$$\phi_n(X) = \frac{d}{dt}\Phi(e^t X)|_{t=0} \Rightarrow \phi_n(X)P(z) = \frac{d}{dt}P(e^{-tX}z)|_{t=0}, \quad z = \begin{pmatrix} z_1 \\ z_2 \end{pmatrix}.$$

Defining $z(t) = e^{-tX}z = (z_1(t), z_2(t))^{\mathrm{T}}$ and using the chain rule one finds

$$\phi_n(X)P = \frac{\partial P}{\partial z_1}\frac{dz_1}{dt}|_{t=0} + \frac{\partial P}{\partial z_2}\frac{dz_2}{dt}|_{t=0} = -\frac{\partial P}{\partial z_1}(X_{11}z_1 + X_{12}z_2) - \frac{\partial P}{\partial z_2}(X_{21}z_1 + X_{22}z_2).$$

The basis elements of **sl**(2, **C**) are then represented by

$$\phi_n(H) = -z_1\frac{\partial}{\partial z_1} + z_2\frac{\partial}{\partial z_2}, \quad \phi_n(X) = -z_2\frac{\partial}{\partial z_1}, \quad \phi_n(Y) = -z_1\frac{\partial}{\partial z_2}.$$

on the space $P \in \mathbb{P}^2 n$ (all n). By employing the unitarian trick one obtains representations for SU(2), **su**(2), SL(2, **R**, and **sl**(2, **R**), all are obtained by restriction of either **(S2)** or **(S4)**. They are formally identical to **(S2)** or **(S4)**. With a choice of basis for $P \in \mathbb{P}^2 n$, all these representations become matrix groups or matrix Lie algebras.

The (μ, ν)-representations are realized on a space of polynomials $\mathbb{P}^2 \mu\nu$ in z_1, $\bar{z}_1$, z_2, $\bar{z}_2$, homogeneous of degree μ in z_1, z_2 and homogeneous of degree ν in $\bar{z}_1$, $\bar{z}_2$.[48] The representations are given by

$$[\Phi_{\mu,\nu}(g)]P(z) = \left[\Phi\begin{pmatrix} a & b \\ c & d \end{pmatrix}P\right]\begin{pmatrix} z_1 \\ z_2 \end{pmatrix} = P\left(\begin{pmatrix} a & b \\ c & d \end{pmatrix}^{-1}\begin{pmatrix} z_1 \\ z_2 \end{pmatrix}\right), \quad P \in \mathbb{P}^2_{\mu,\nu}.$$

By carrying out the same steps as above, one finds

$$\phi_{\mu,\nu}(X)P = -\frac{\partial P}{\partial z_1}(X_{11}z_1 + X_{12}z_2) - \frac{\partial P}{\partial z_2}(X_{21}z_1 + X_{22}z_2) - \frac{\partial P}{\partial \bar{z}_1}(\overline{X_{11}\bar{z}_1} + \overline{X_{12}\bar{z}_2}) - \frac{\partial P}{\partial \bar{z}_2}(\overline{X_{21}\bar{z}_1} + \overline{X_{22}\bar{z}_2}), \quad X \in \mathfrak{sl}(2,\mathbb{C}),$$

from which the expressions

$$\phi_{\mu,\nu}(H) = -z_1\frac{\partial}{\partial z_1} + z_2\frac{\partial}{\partial z_2} - \bar{z}_1\frac{\partial}{\partial \bar{z}_1} + \bar{z}_2\frac{\partial}{\partial \bar{z}_2}, \quad \phi_{\mu,\nu}(X) = -z_2\frac{\partial}{\partial z_1} - \bar{z}_2\frac{\partial}{\partial \bar{z}_1}, \quad \phi_{\mu,\nu}(Y) = -z_1\frac{\partial}{\partial z_2} - \bar{z}_1\frac{\partial}{\partial \bar{z}_2}$$

for the basis elements follow.

Non-surjectiveness of exponential mapping   Unlike in the case exp: **so**(3; 1) → SO(3; 1)⁺, the exponential mapping exp: **sl**(2, **C**) → SL(2, **C**) is not onto.[55] The conjugacy classes of SL(2, **C**) are represented by the matrices

$$a(\alpha) = \begin{pmatrix} \alpha & 0 \\ 0 & \alpha^{-1} \end{pmatrix} = e^{\log(\alpha)H}, \alpha \neq 0, \quad p = \begin{pmatrix} 1 & 1 \\ 0 & 1 \end{pmatrix} = e^X, \quad q = \begin{pmatrix} -1 & 1 \\ 0 & -1 \end{pmatrix},$$

but there is no element Q in **sl**(2, **C**) such that $q = \exp(Q)$.[nb 9]

In general, if g is an element of a connected Lie group G with Lie algebra **g**, then

$$g = e^{X_1}e^{X_2}\cdots e^{X_n}, \quad X_i \in \mathfrak{g}, 1 \leq i \leq n.$$

This follows from the compactness of a path from the identity to g and the one-to-one nature of exp near the identity. In the case of the matrix q, one may write

The kernel of the covering map $\mathbf{p}$:$SL(2, \mathbf{C}) \to \mathbf{SO}(3; 1)^+$ of above is $N = \{I, -I\}$, a normal subgroup of $SL(2, \mathbf{C})^+$. The composition $\mathbf{p} \circ \exp$: $\mathbf{sl}(2, \mathbf{C}) \to SO(3; 1)$ *is* onto. If a matrix a is not in the image of exp, then there is a matrix b equivalent to it with respect to $\mathbf{p}$, meaning $\mathbf{p}(b) = \mathbf{p}(a)$, that is in the image of exp. The condition for equivalence is $a^{-1}b \in N$.[58] In the case of the matrix q, one may solve for p in the equation $p^{-1}q = -I \in N$. One finds

$$ p = \begin{pmatrix} 1 & -1 \\ 0 & 1 \end{pmatrix} = e^{-X}, \quad p^{-1}q = \begin{pmatrix} 1 & 1 \\ 0 & 1 \end{pmatrix} \begin{pmatrix} -1 & 1 \\ 0 & -1 \end{pmatrix} = \begin{pmatrix} -1 & 0 \\ 0 & -1 \end{pmatrix} = -I. $$

As a corollary, since the covering map $\mathbf{p}$ is a homomorphism, the mapping version of the Lie correspondence **(G6)** can be used to provide a proof of the surjectiveness of exp for $\mathbf{so}(3; 1)$. Let σ denote the isomorphism between $\mathbf{sl}(2, \mathbf{C})$ and $\mathbf{so}(3; 1)$. Refer to the commutative diagram. One has $\mathbf{p} \circ \exp$: $\mathbf{sl}(2, \mathbf{C}) \to SO(3; 1) = \exp \circ \sigma$ for all $X \in \mathbf{sl}(2, \mathbf{C})$. Since $\mathbf{p} \circ \exp$ is onto, $\exp \circ \sigma$ is onto, and hence $\exp$: $\mathbf{so}(3; 1) \to SO(3; 1)^+$ is onto as well.

SO(3; 1)$^+$-representations from SL(2, C)-representations   By the first isomorphism theorem, a representation (Φ, V) of $SL(2, \mathbf{C})$ descends to a representation (Π, V) of $SO(3; 1)^+$ if and only if ker $\mathbf{p} \subset$ ker Φ. Refer to the commutative diagram. If this condition holds, then both elements in the fiber $\mathbf{p}^{-1}(g)$, $g \in SO(3; 1)^+$ will be mapped by Φ to the same representative, and the expression $\Phi(\mathbf{p}^{-1}(g))$ makes sense. One may thus define Π: $SO(3; 1)^+ \to GL(V)$, $\Pi(g) = \Phi(\mathbf{p}^{-1}(g))$. In particular, if Π is faithful, i.e. having kernel $= I$, then there is no corresponding proper representation of $SO(3; 1)^+$, but there is a projective one as was shown in a previous section, corresponding to the two possible choices of representative in each fiber $\mathbf{p}^{-1}(g)$.

Lie algebra representations of $\mathbf{so}(3; 1)$ are obtained from $\mathbf{sl}(2, \mathbf{C})$-representations simply by composition with σ^{-1}.

SL(2, C)-representations from SO(3; 1)$^+$-representations $SL(2, \mathbf{C})$-representations can be obtained from non-projective $SO(3; 1)^+$-representations by composition with the projection map $\mathbf{p}$. These are always representations since they are compositions of group homomorphisms. Such a representation is never faithful because Ker $\mathbf{p} = \{I, -I\}$. If the $SO(3; 1)^+$-representation is projective, then the resulting $SL(2, \mathbf{C})$-representation would be projective as well. Instead, the isomorphism σ:$\mathbf{so}(3; 1) \to \mathbf{sl}(3, \mathbf{C})$ can be employed, composed with $\exp$:$\mathbf{sl}(2, \mathbf{C}) \to SL(2, \mathbf{C})$. This is always a non-projective representation.

11.1.6   Properties of the (m, n) representations

The (m, n) representations are $(2m + 1)(2n + 1)$-dimensional,[59] irreducible, and they are the only irreducible representations.[27]

- Irreducibility follows from the unitarian trick[26] and that a representation Π of $SU(2) \times SU(2)$ is irreducible if and only if $\Pi = \Pi\mu \otimes \Pi\nu$,[nb 10] where $\Pi\mu$, $\Pi\nu$ are irreducible representations of $SU(2)$.

- Uniqueness follows from that the Πm are the only irreducible representations of $SU(2)$, which is one of the conclusions of the theorem of the highest weight.[60]

- The dimensionality follows from the Weyl dimension formula. For a Lie algebra $\mathbf{g}$ it reads

$$ \dim \pi_\mu = \frac{\prod_{\alpha \in R^+} \langle \alpha, \mu + \delta \rangle}{\prod_{\alpha \in R^+} \langle \alpha, \delta \rangle}, \quad \text{[61]} $$

where R^+ is the set of positive roots and δ is half the sum of the positive roots. The inner product $\langle \cdot, \cdot \rangle$ is that of the Lie algebra $\mathbf{g}$, invariant under the action of the Weyl group on $\mathbf{h} \subset \mathbf{g}$, the Cartan subalgebra. The roots (really elements of $\mathbf{h}^*$) are via this inner product identified with elements of $\mathbf{h}$. For $\mathbf{sl}(2, \mathbf{C})$, the formula reduces to $\dim \pi\mu = \mu + 1 = 2m + 1$.[62] By taking tensor products, the result follows. A quicker approach is, of course, to simply count the dimensions in any concrete realization, such as the one given in Representations of $SL(2, \mathbf{C})$ and $\mathbf{sl}(2, \mathbf{C})$.

Faithfulness

If a representation Π of a Lie group G is not faithful, then $N = \ker \Pi$ is a nontrivial normal subgroup because $\Pi(n) = I \Rightarrow \Pi(gng^{-1}) = \Pi(g)\Pi(n)\Pi(g)^{-1} = \Pi(g)\Pi(g)^{-1} = I$. There are three relevant cases.

1. N is non-discrete and abelian.

2. N is non-discrete and non-abelian.

3. N is discrete. In this case $N \subset Z$, where Z is the center of G.[nb 11]

In the case of SO(3; 1)$^+$, the first case is excluded since SO(3; 1)$^+$ is semi-simple.[nb 12] The second case (and the first case) is excluded because SO(3; 1)$^+$ is simple.[nb 13] For the third case, SO(3; 1)$^+$ is isomorphic to the quotient SL(2, **C**)/{I, $-I$}. But {I, $-I$} is the center of SL(2, **C**). It follows that the center of SO(3; 1)$^+$ is trivial, and this excludes the third case. The conclusion is that every representation Π:SO(3; 1)$^+ \to$ GL(V) and every projective representation Π:SO(3; 1)$^+ \to$ PGL(W) for V, W finite-dimensional vector spaces are faithful.

By using the fundamental Lie correspondence, the statements and the reasoning above translate directly to Lie algebras with (abelian) nontrivial non-discrete normal subgroups replaced by (one-dimensional) nontrivial ideals in the Lie algebra,[63] and the center of SO(3; 1)$^+$ replaced by the center of **sl**(3; 1)$^+$. The center of any semisimple Lie algebra is trivial[64] and **so**(3; 1) is semi-simple and simple, and hence has no non-trivial ideals.

Non-unitarity

The (m, n) Lie algebra representation is not Hermitian. Accordingly, the corresponding (projective) representation of the group is never unitary. This is due to the non-compactness of the Lorentz group. A connected simple non-compact Lie group cannot have *any* nontrivial finite-dimensional irreducible representations.[65] There is a topological proof of this.[66] Let U:$G \to$ GL(V), where V is finite-dimensional, be a continuous unitary representation of the non-compact connected simple Lie group G. Then $U(G) \subset$ U(V) $\subset$ GL(V) where U(V) is the compact subgroup of GL(V) consisting of unitary transformations of v). The kernel, $\ker U$, of U is a normal subgroup of G. Since G is simple, $\ker U$ is either all of G, in which case U is trivial, or $\ker U$ is trivial, in which case U is faithful. In the latter case U is a diffeomorphism onto its image,[67] $U(G) \approx G$., and $U(G)$ is Lie group. This would mean that $U(G)$ is an embedded non-compact Lie subgroup of the compact group U(V). This is impossible with the subspace topology on $U(G) \subset$ U(V) since all *embedded* Lie subgroups of a Lie group are closed[68] If $U(G)$ were closed, it would be compact,[nb 14] and the G would be compact,[nb 15] contrary to assumption.

In the case of the Lorentz group, this can also be seen directly from the definitions. The representations of **A** and **B** used in the construction are Hermitian. This means that **J** is Hermitian, but **K** is anti-Hermitian.[28] The non-unitarity is not a problem in quantum field theory, since the objects of concern are not required to have a Lorentz-invariant positive definite norm.[28]

Restriction to SO(3)

The (m, n) representation is, however, unitary when restricted to the rotation subgroup SO(3), but these representations are not irreducible as representations of SO(3). A Clebsch–Gordan decomposition can be applied showing that an (m, n) representation have SO(3)-invariant subspaces of highest weight (spin) $m + n$, $m + n - 1$, ... , | $m - n$ |,[28] where each possible highest weight (spin) occurs exactly once. A weight subspace of highest weight (spin) j is $(2j + 1)$-dimensional. So for example, the (1/2, 1/2) representation has spin 1 and spin 0 subspaces of dimension 3 and 1 respectively.

Since the angular momentum operator is given by **J** = **A** + **B**, the highest spin in quantum mechanics of the rotation sub-representation will be $(m + n)\hbar$ and the "usual" rules of addition of angular momenta and the formalism of 3-j symbols, 6-j symbols, etc. applies.[69]

Spinors

It is the SO(3)-invariant subspaces subspaces of the irreducible representations that determine whether a representation has spin. From the above paragraph, it is seen that the (m, n) representation has spin if $m + n$ is half-integral. The simplest are (1/2, 0) and (0, 1/2), the Weyl-spinors of dimension 2. Then, for example, (0, 3/2) and (1, 1/2) are a spin representations of dimensions 23/2 + 1 = 4 and (2 + 1)(21/2 + 1) = 6 respectively. Note that, according to the above paragraph, there are subspaces with spin both 3/2 and 1/2 in the last two cases, so these representations cannot likely represent a *single* physical particle which must be well-behaved under SO(3).

Construction of pure spin $n/2$ representations for any n (under SO(3)) from the irreducible representations involves taking tensor products of the Dirac-representation with a non-spin representation, extraction of a suitable subspace, and finally imposing differential constraints.[70]

Dual representations

To see if the dual representation of an irreducible representation is equivalent to the original representation one can consider the following theorems:

1. The set of weights of the dual representation of an irreducible representation of a semisimple Lie algebra is, including multiplicities, the negative of the set of weights for the original representation.[71]

2. Two irreducible representations are equivalent if and only if they have the same highest weight.[nb 16]

3. For each semisimple Lie algebra there exists a unique element w_0 of the Weyl group such that if μ is a dominant integral weight, then $w_0 \cdot (-\mu)$ is again a dominant integral weight.[72]

4. If $\pi\mu_0$ is an irreducible representation with highest weight μ_0, then $\pi^*\mu_0$ has highest weight $w_0 \cdot (-\mu)$.[73]

Here, the elements of the Weyl group are considered as orthogonal transformations, acting by matrix multiplication, on the real vector space of roots. One sees that if $-I$ is an element of the Weyl group of a semisimple Lie algebra, then $w_0 = -I$. In the case of $\mathbf{sl}(2, \mathbf{C})$, the Weyl group is $W = \{I, -I\}$.[74] It follows that each $\pi\mu$, $\mu = 0, 1, \ldots$ is equivalent to its dual $\pi\mu^*$. The root system of $\mathbf{sl}(2, \mathbf{C}) \oplus \mathbf{sl}(2, \mathbf{C})$ is shown in the figure to the right.[nb 17] The Weyl group is generated by $\{w_\gamma\}$ where w_γ is reflection in the plane orthogonal to γ as γ ranges over all roots.[nb 18] One sees that $w\alpha \cdot w_\beta = -I$ so $-I \in W$. Then using the fact that if π, σ are Lie algebra representations and $\pi \approx \sigma$, then $\Pi \approx \Sigma$.[75] The conclusion for SO(3; 1)$^+$ is

$$\pi^*_{m,n} \cong \pi_{m,n}, \quad \Pi^*_{m,n} \cong \Pi_{m,n}, \quad 2m, 2n \in \mathbb{N}.$$

Complex conjugate representations

If π is a representation of a Lie algebra, then $\overline{\pi}$ is a representation, where the bar denotes entry-wise complex conjugation in the representative matrices. This follows from that complex conjugation commutes with addition and multiplication.[76] In general, every irreducible representation π of $\mathbf{sl}(n, \mathbf{C})$ can be written uniquely as $\pi = \pi^+ + \pi^-$, where

$$\pi^\pm(X) = \tfrac{1}{2}(\pi(X) \pm i\pi(i^{-1}X)), \quad [77]$$

with π^+ holomorphic (complex linear) and π^- **anti-holomorphic** (conjugate linear). For $\mathbf{sl}(2, \mathbf{C})$, since $\pi\mu$ is holomorphic, $\overline{\pi\mu}$ is anti-holomorphic. Direct examination of the explicit expressions for $\pi\mu$, $_0$ and π_0, ν in equation **(S8)** below shows that they are holomorphic and anti-holomorphic respectively. Closer examination of the expression **(S8)** also allows for identification of π^+ and π^- for $\pi\mu$, ν as $\pi^+\mu$, $\nu = \pi\mu^{\oplus\nu + 1}$ and $\pi^-\mu$, $\nu = \pi\nu^{\oplus\mu + 1}$.

Using the above identities (interpreted as pointwise addition of functions), for SO(3; 1)$^+$ yields

$$\overline{\pi_{m,n}} = \overline{\pi^+_{m,n} + \pi^-_{m,n}} = \overline{\pi_m^{\oplus 2n+1}} + \overline{\pi_n^{\oplus 2m+1}} = \pi_n^{\oplus 2m+1} + \overline{\pi_m}^{\oplus 2n+1} = \pi^+_{n,m} + \pi^-_{n,m} = \pi_{n,m}, \quad \overline{\Pi_{m,n}} = \Pi_{n,m}, \quad 2m, 2n \in \mathbb{N},$$

where the statement for the group representations follow from exp(X) = exp(X). It follows that the irreducible representations (m, n) have real matrix representatives if and only if $m = n$. Reducible representations on the form $(m, n) \oplus (n, m)$ have real matrices too.

11.1.7 Induced representations

In general representation theory, if (π, V) is a representation of a Lie algebra $\mathbf{g}$, then there is an associated representation of $\mathbf{g}$ on End V, also denoted π, given by

$$\pi(X)(A) = [\pi(X), A], \quad A \in \text{End } V, \ X \in \mathfrak{g}.$$

Likewise, a representation (Π, V) of a group G yields a representation Π on End V of G, still denoted Π, given by

$$\Pi(g)(A) = \Pi(g)A\Pi(g)^{-1}, \quad A \in \text{End } V, \ g \in G.$$

Applying this to the Lorentz group, if (Π, V) is a projective representation, then direct calculation using (G4) shows that the induced representation on End V is, in fact, a proper representation, i.e. a representation without phase factors.

In quantum mechanics this means that if (π, H) or (Π, H) is a representation acting on some Hilbert space H, then the corresponding induced representation acts on the set of linear operators on H. As an example, the induced representation of the projective spin $(1/2, 0) \oplus (0, 1/2)$ representation on End(H) is the non-projective 4-vector $(1/2, 1/2)$ representation.[28]

For simplicity, consider now only the "discrete part" of End H, that is, given a basis for H, the set of constant matrices of various dimension, including possibly infinite dimensions. A general element of the full End H is the sum of tensor products of a matrix from the simplified End H and an operator from the left out part. The left out part consists of functions of spacetime, differential and integral operators and the like. See Dirac operator for an illustrative example. Also left out are operators corresponding to other degrees of freedom not related to spacetime, such as gauge degrees of freedom in gauge theories.

The induced 4-vector representation of above on this simplified End H has an invariant 4-dimensional subspace that is spanned by the four gamma matrices.[28] (Note the different metric convention in the linked article.) In a corresponding way, the complete Clifford algebra of spacetime, $C\ell_{3,1}(\mathbf{R})$, whose complexification is $M_4(\mathbf{C})$, generated by the gamma matrices decomposes as a direct sum of representation spaces of a **scalar** irreducible representation (irrep), the (0, 0), a **pseudoscalar** irrep, also the (0, 0), but with parity inversion eigenvalue −1, see the next section below, the already mentioned **vector** irrep, (1/2, ,1/2), a **pseudovector** irrep, (1/2, 1/2) with parity inversion eigenvalue +1 (not −1), and a **tensor** irrep, (1, 0) $\oplus$ (0, 1).[28] The dimensions add up to $1 + 1 + 4 + 4 + 6 = 16$. In other words,

$$Cl_{3,1}(\mathbb{R}) = (0,0) \oplus \left(\frac{1}{2}, \frac{1}{2}\right) \oplus [(1,0) \oplus (0,1)] \oplus \left(\frac{1}{2}, \frac{1}{2}\right)_p \oplus (0,0)_p,$$

where, as is customary, a representation is confused with its representation space. This is, in fact, a reasonably convenient way to show that the algebra spanned by the gammas is 16-dimensional.[79]

The six-dimensional representation space of the tensor (1, 0) $\oplus$ (0, 1)-representation inside $C\ell_{3,1}(\mathbf{R})$ has two roles. In particular, letting

$$\sigma^{\mu\nu} = -\frac{i}{4}[\gamma^\mu, \gamma^\nu],$$

where $\{\gamma^\mu \in C\ell_{3,1}(\mathbf{R}): \mu = 0,1,2,3\}$ are the gamma matrices, the $\{\sigma^{\mu\nu} \in C\ell_{3,1}(\mathbf{R})\}$, only 6 of which are non-zero due to antisymmetry of the bracket, span the tensor representation space. Moreover, they have the commutation relations of the Lorentz Lie algebra,

$$[\sigma^{\mu\nu}, \sigma^{\rho\tau}] = i(\eta^{\tau\mu}\sigma^{\rho\nu} + \eta^{\nu\tau}\sigma^{\mu\rho} - \eta^{\rho\mu}\sigma^{\tau\nu} - \eta^{\nu\rho}\sigma^{\mu\tau}),$$

and hence constitute a representation (in addition to being a representation space) sitting inside $C\ell_{3,1}(\mathbf{R})$, the (1/2, 0) $\oplus$ (0, 1/2) spin representation. For details, see bispinor and Dirac algebra.

The conclusion is that every element of the complexified $C\ell_{3,1}(\mathbf{R})$ in End H (i.e. every complex 4×4 matrix) has well defined Lorentz transformation properties. In addition, it has a spin-representation of the Lorentz Lie algebra, which upon exponentiation becomes a spin representation of the group, acting on $\mathbf{C}^4$, making it a space of bispinors.

There is also a multitude of other representations that can be said being "induced" by the irreducible ones, such as those obtained in a standard manner by taking direct sums, tensor products, dual representations, quotients, etc. of the irreducible representations. These are not discussed here.

11.1.8   The full Lorentz group

The (possibly projective) (m, n) representation is irreducible as a representation SO(3; 1)$^+$, the identity component of the Lorentz group, in physics terminology the proper orthochronous Lorentz group. If $m = n$ it can be extended to a representation of all of SO(3; 1), the full Lorentz group, including space parity inversion and time reversal.[28]

Space parity inversion

For space parity inversion, one considers the adjoint action AdP of $P \in$ SO(3; 1) on $\mathbf{so}(3; 1)$, where P is the standard representative of space parity inversion, $P = \mathrm{diag}(1, -1, -1, -1)$, given by

$$\mathrm{Ad}_P(J_i) = PJ_iP^{-1} = J_i, \qquad \mathrm{Ad}_P(K_i) = PK_iP^{-1} = -K_i.$$

It is these properties of $\mathbf{K}$ and $\mathbf{J}$ under P that motivate the terms *vector* for $\mathbf{K}$ and pseudovector or *axial vector* for $\mathbf{J}$. In a similar way, if π is any representation of $\mathbf{so}(3; 1)$ and Π is its associated group representation, then Π(SO(3; 1)$^+$) acts on the representation of π by the adjoint action, $\pi(X) \mapsto \Pi(g)\,\pi(X)\,\Pi(g)^{-1}$ for $X \in \mathbf{so}(3; 1)$, g $\in$ SO(3; 1)$^+$. If P is to be included in Π, then consistency with **(F1)** requires that

$$\Pi(P)\pi(B_i)\Pi(P)^{-1} = \pi(A_i)$$

holds, where $\mathbf{A}$ and $\mathbf{B}$ are defined as in the first section. This can hold only if Ai and Bi have the same dimensions, i.e. only if $m = n$. When $m \neq n$ then $(m, n) \oplus (n, m)$ can be extended to an irreducible representation of SO(3; 1)$^+$, the orthocronous Lorentz group. The parity reversal representative $\Pi(P)$ does not come automatically with the general construction of the (m, n) representations. It must be specified separately. The matrix $\beta = i\,\gamma^0$ (or a multiple of modulus -1 times it) may be used in the (1/2, 0) $\oplus$ (0, 1/2)[28] representation.

If parity is included with a minus sign (the 1×1 matrix $[-1]$) in the (0,0) representation, it is called a pseudoscalar representation.

Time reversal

Time reversal $T = \mathrm{diag}(-1, 1, 1, 1)$, acts similarly on $\mathbf{so}(3; 1)$ by

$$\mathrm{Ad}_T(J_i) = TJ_iT^{-1} = -J_i, \qquad \mathrm{Ad}_T(K_i) = TK_iT^{-1} = K_i.$$

By explicitly including a representative for T, as well as one for P, one obtains a representation of the full Lorentz group SO(3; 1). A subtle problem appears however in application to physics, in particular quantum mechanics. When considering the full Poincaré group, four more generators, the P^μ, in addition to the J^i and K^i generate the group. These are interpreted as generators of translations. The time-component P^0 is the Hamiltonian H. The operator T satisfies the relation

$$\mathrm{Ad}_T(iH) = TiHT^{-1} = -iH$$

in analogy to the relations above with **so**(3; 1) replaced by the full Poincaré algebra. By just cancelling the i's, the result $THT^{-1} = -H$ would imply that for every state Ψ with positive energy E in a Hilbert space of quantum states with time-reversal invariance, there would be a state $\Pi(T^{-1})\Psi$ with negative energy $-E$. Such states do not exist. The operator $\Pi(T)$ is therefore chosen antilinear and antiunitary, so that it anticommutes with i, resulting in $THT^{-1} = +H$, and its action on Hilbert space likewise becomes antilinear and antiunitary.[83] It may be expressed as the composition of complex conjugation with multiplication by a unitary matrix.[84][85] This is mathematically sound, see Wigner's theorem, but if one is very strict with terminology, Π is not a *representation*.

When constructing theories such as QED which is invariant under space parity and time reversal, Dirac spinors may be used, while theories that do not, such as the electroweak force, must be formulated in terms of Weyl spinors. The Dirac representation, $(1/2, 0) \oplus (0, 1/2)$, is usually taken to include both space parity and time inversions. Without space parity inversion, it is not an irreducible representation.

The third discrete symmetry entering in the CPT theorem along with P and T, charge conjugation symmetry C, has nothing directly to do with Lorentz invariance.[86]

11.2 Infinite-dimensional representations

11.2.1 History

The Lorentz group $SO(3; 1)^+$ and its double cover $SL(2, \mathbf{C})$ also have infinite dimensional unitary representations, first studied independently by Bargmann (1947), Gelfand & Naimark (1947) and Harish-Chandra (1947) at the instigation of Paul Dirac. This trail of development begun with Dirac (1936) where he devised matrices **U** and **B** necessary for description of higher spin (compare Dirac matrices), elaborated upon by Fierz (1939), see also Fierz & Pauli (1939), and proposed precursors of the Bargmann-Wigner equations. In Dirac (1945) he proposed a concrete infinite-dimensional representation space whose elements were called **expansors** as a generalization of tensors. These ideas were incorporated by Harish–Chandra and expanded with **expinors** as an infinite-dimensional generalization of spinors in his 1947 paper.

The Plancherel formula for these groups was first obtained by Gelfand and Naimark through involved calculations. The treatment was subsequently considerably simplified by Harish-Chandra (1951) and Gelfand & Graev (1953), based on an analogue for $SL(2, \mathbf{C})$ of the integration formula of Hermann Weyl for compact Lie groups. Elementary accounts of this approach can be found in Rühl (1970) and Knapp (2001).

The theory of spherical functions for the Lorentz group, required for harmonic analysis on the 3-dimensional unit quasi-sphere in Minkowski space, or equivalently 3-dimensional hyperbolic space, is considerably easier than the general theory. It only involves representations from the spherical principal series and can be treated directly, because in radial coordinates the Laplacian on the hyperboloid is equivalent to the Laplacian on **R**. This theory is discussed in Takahashi (1963), Helgason (1968), Helgason (2000) and the posthumous text of Jorgenson & Lang (2008).

11.2.2 Action on function spaces

In the classification of the irreducible finite-dimensional representations of above it was never specified precisely *how* a representative of a group or Lie algebra element acts on vectors in the representation space. The action can be anything as long as it is linear. The point silently adopted was that after a choice of basis in the representation space, everything becomes matrices anyway.

If V is a vector space of functions of a finite number of variables n, then the action on a scalar function $f \in V$ given by

$$(\Pi(g)f)(x) = f(\Pi_x(g)^{-1}x), \qquad x \in \mathbb{R}^n, f \in V$$

produces another function $\Pi f \in V$. Here Πx is an n-dimensional representation, and Π is a possibly infinite-dimensional representation. A special case of this construction is when V is a space of functions defined on the group G itself, viewed as a n-dimensional manifold embedded in $\mathbf{R}^n$.[87] This is the setting in which the Peter–Weyl theorem and the Borel–Weil theorem are formulated. The former demonstrates the existence of a Fourier decomposition of functions on a compact group into characters of finite-dimensional representations.[27] The completeness of the characters in this sense can thus be used to prove the existence of the highest weight representations.[22] The latter theorem, providing more explicit representations, makes use of the unitarian trick to yield representations of complex non-compact groups, e.g. SL(2, **C**); in the present case, there is a one-to-one correspondence between representations of SU(2) and holomorphic representations of SL(2, **C**). (A group representation is called holomorphic if its corresponding Lie algebra representation is complex linear.) This theorem too can be used to demonstrate the existence of the highest weight representations.[22]

Euclidean rotations

Main articles: Rotation group SO(3) and Spherical harmonics

The subgroup SO(3) of three-dimensional Euclidean rotations has an infinite-dimensional representation on the Hilbert space $L^2(\mathbf{S}^2) = \mathrm{span}\{Y^\ell m, \ell \in \mathbf{N}^+, -\ell \le m \le \ell \}$, where the $Y^\ell m$ are spherical harmonics. Its elements are square integrable complex-valued functions[nb 19] on the sphere. The inner product on this space is given by

$$\langle f, g \rangle = \int_{\mathbb{S}^2} \overline{f} g \, d\Omega = \int_0^{2\pi} \int_0^\pi \overline{f} g \sin\theta \, d\theta \, d\varphi.$$

If f is an arbitrary square integrable function defined on the unit sphere $\mathbf{S}^2$, then it can be expressed as[88]

$$|f\rangle = \sum_{l=1}^\infty \sum_{m=-l}^{m=l} |Y_m^l\rangle\langle Y_m^l|f\rangle, \quad f(\theta, \varphi) = \sum_{l=1}^\infty \sum_{m=-l}^{m=l} f_{lm} Y_m^l(\theta, \varphi),$$

where the expansion coefficients are given by

$$f_{lm} = \langle Y_m^l, f \rangle = \int_{\mathbb{S}^2} \overline{Y_m^l} f \, d\Omega = \int_0^{2\pi} \int_0^\pi \overline{Y_m^l}(\theta, \varphi) f(\theta, \varphi) \sin\theta \, d\theta \, d\varphi.$$

The Lorentz group action restricts to that of SO(3) and is expressed as

$$(\Pi(R)f)(\theta(x), \varphi(x)) = \sum_{l=1}^\infty \sum_{m=-l}^{m=l} \sum_{m'=-l}^{m'=l} D_{mm'}^{(l)}(R) f_{lm'} Y_m^l(\theta(R^{-1}x), \varphi(R^{-1}x)), \qquad R \in \mathrm{SO}(3), \quad x \in \mathbb{S}^2.$$

This action is unitary, meaning that

$$\langle \Pi(R)f, \Pi(R)g \rangle = \langle f, g \rangle \qquad \forall f, g \in \mathbb{S}^2, \quad \forall R \in \mathrm{SO}(3).$$

The $D^{(\ell)}$ can be obtained from the $D^{(m, n)}$ of above using Clebsch–Gordan decomposition, but they are more easily directly expressed as an exponential of an odd-dimensional $\mathbf{su}(2)$-representation (the 3-dimensional one is exactly $\mathbf{so}(3)$).[89][90] In this case the space $L^2(\mathbf{S}^2)$ decomposes neatly into an infinite direct sum of irreducible odd finite-dimensional representations V_{2i+1}, $i = 0, 1, \ldots$ according to

$$L^2(\mathbb{S}^2) = \sum_{i=0}^\infty V_{2i+1} \equiv \bigoplus_{i=0}^\infty \mathrm{span}\{Y_m^{2i+1}\}$$

This is characteristic of infinite-dimensional unitary representations of SO(3). If Π is an infinite-dimensional unitary representation on a separable[nb 20] Hilbert space, then it decomposes as a direct sum of finite-dimensional unitary representations.[88] Such a representation is thus never irreducible. All irreducible finite-dimensional representations (Π, V) can be made unitary by an appropriate choice of inner product,[88]

$$\langle f, g \rangle_U \equiv \int_{\mathrm{SO}(3)} \langle \Pi(R)f, \Pi(R)g \rangle dg = \frac{1}{8\pi^2} \int_0^{2\pi} \int_0^\pi \int_0^{2\pi} \langle \Pi(R)f, \Pi(R)g \rangle \sin\theta \, d\varphi \, d\theta \, d\psi, \quad f, g \in V,$$

where the integral is the unique invariant integral over SO(3) normalized to 1, here expressed using the Euler angles parametrization. The inner product inside the integral is any inner product on V.

11.2.3 The Möbius group

Main article: Möbius transformation

The identity component of the Lorentz group is isomorphic to the Möbius group M as is described in detail in Lorentz group. This group can be thought of as conformal mappings of either the complex plane or, via stereographic projection, the Riemann sphere. In this way, the Lorentz group itself can be thought of as acting conformally on the complex plane or on the Riemann sphere. In the plane, a Möbius transformation characterized by the complex numbers a, b, c, d acts on the plane according to

$$f(z) = \frac{az + b}{cz + d}, \qquad ad - bc \neq 0$$

and can be represented by complex matrices

$$\Pi_f = \begin{pmatrix} a & b \\ c & d \end{pmatrix}, \qquad \det \Pi_f = 1.$$

These are elements of SL(2, **C**) and are unique up to a sign and M $\approx$ SL(2, **C**)/{I, –I} $\approx$ SO(3; 1)$^+$. The conformal mappings of the Riemann sphere are thoroughly described in Möbius transformations.

11.2.4 The Riemann P-functions

Main article: Riemann's differential equation

The Riemann P-functions are an example of a set of functions that transform among themselves under the action of the Lorentz (Möbius) group. The Riemann P-functions are expressed as

$$w(z) = P \left\{ \begin{matrix} a & b & c & \\ \alpha & \beta & \gamma & z \\ \alpha' & \beta' & \gamma' & \end{matrix} \right\} = \left(\frac{z-a}{z-b} \right)^{\alpha} \left(\frac{z-c}{z-b} \right)^{\gamma} P \left\{ \begin{matrix} 0 & \infty & 1 & \\ 0 & \alpha+\beta+\gamma & 0 & \frac{(z-a)(c-b)}{(z-b)(c-a)} \\ \alpha'-\alpha & \alpha+\beta'+\gamma & \gamma'-\gamma & \end{matrix} \right\}$$

where the $a, b, c, \alpha, \beta, \gamma, \alpha', \beta', \gamma'$ are complex constants. The P-function on the right hand side can be expressed using standard hypergeometric functions giving

$$w(z) = \left(\frac{z-a}{z-b} \right)^{\alpha} \left(\frac{z-c}{z-b} \right)^{\gamma} {}_2F_1 \left(\alpha+\beta+\gamma, \alpha+\beta'+\gamma; 1+\alpha-\alpha'; \frac{(z-a)(c-b)}{(z-b)(c-a)} \right).$$

Now define an action of the Lorentz group on the set of all Riemann P-functions by

$$u = \frac{Az + B}{Cz + D} \quad \text{and} \quad \eta = \frac{Aa + B}{Ca + D}$$

and

$$\zeta = \frac{Ab + B}{Cb + D} \quad \text{and} \quad \theta = \frac{Ac + B}{Cc + D},$$

where A, B, C, D are the entries in

$$\pi_{f_\Lambda}^{-1} = \begin{pmatrix} A & B \\ C & D \end{pmatrix},$$

where, Λ is a Lorentz transformation and $f\Lambda$ is the corresponding Möbius transformation, and, finally, $\Pi f\Lambda$ is one of the two possible SL(2, **C**) matrices corresponding to it, then one has the relation

$$P \left\{ \begin{matrix} a & b & c & \\ \alpha & \beta & \gamma & z \\ \alpha' & \beta' & \gamma' & \end{matrix} \right\} = P \left\{ \begin{matrix} \eta & \zeta & \theta & \\ \alpha & \beta & \gamma & u \\ \alpha' & \beta' & \gamma' & \end{matrix} \right\}$$

expressing the symmetry. The inverse in **T5** is needed to obtain a (local) homomorphism.

11.2.5   Principal series

The **principal series**, or **unitary principal series**, are the unitary representations induced from the one-dimensional representations of the lower triangular subgroup B of $G = SL(2, \mathbf{C})$. Since the one-dimensional representations of B correspond to the representations of the diagonal matrices, with non-zero complex entries z and z^{-1}, they thus have the form

$$\chi_{\nu,k}\begin{pmatrix} z & 0 \\ c & z^{-1} \end{pmatrix} = r^{i\nu}e^{ik\theta},$$

for k an integer, ν real and with $z = re^{i\theta}$. The representations are irreducible; the only repetitions occur when k is replaced by $-k$. By definition the representations are realized on L^2 sections of line bundles on $G/B = S^2$, which is isomorphic to the Riemann sphere. When $k = 0$, these representations constitute the so-called **spherical principal series**.

The restriction of a principal series to the maximal compact subgroup $K = SU(2)$ of G can also be realized as an induced representation of K using the identification $G / B = K / T$, where $T = B \cap K$ is the maximal torus in K consisting of diagonal matrices with $| z | = 1$. It is the representation induced from the 1-dimensional representation z^k T, and is independent of ν. By Frobenius reciprocity, on K they decompose as a direct sum of the irreducible representations of K with dimensions $|k| + 2m + 1$ with m a non-negative integer.

Using the identification between the Riemann sphere minus a point and $\mathbf{C}$, the principal series can be defined directly on $L^2(\mathbf{C})$ by the formula

$$\pi_{\nu,k}\begin{pmatrix} a & b \\ c & d \end{pmatrix}^{-1} f(z) = |cz + d|^{-2-i\nu}\left(\frac{cz+d}{|cz+d|}\right)^{-k} f\left(\frac{az+b}{cz+d}\right). \quad \text{[92]}$$

Irreducibility can be checked in a variety of ways:

- The representation is already irreducible on B. This can be seen directly, but is also a special case of general results on ireducibility of induced representations due to François Bruhat and George Mackey, relying on the Bruhat decomposition $G = B \cup B\, s\, B$ where s is the Weyl group element $\begin{pmatrix} 0 & -1 \\ 1 & 0 \end{pmatrix}$. [93]

- The action of the Lie algebra $\mathfrak{g}$ of G can be computed on the algebraic direct sum of the irreducible subspaces of K can be computed explicitly and the it can be verified directly that the lowest-dimensional subspace generates this direct sum as a $\mathfrak{g}$ -module.[94][95]

11.2.6   Complementary series

The for $0 < t < 2$, the complementary series is defined on L^2 functions f on $\mathbf{C}$ for the inner product

$$(f,g) = \int \int \frac{f(z)\overline{g(w)}\, dz\, dw}{|z-w|^{2-t}}. \quad \text{[96]}$$

with the action given by

$$\pi_t\begin{pmatrix} a & b \\ c & d \end{pmatrix}^{-1} f(z) = |cz + d|^{-2-t} f\left(\frac{az+b}{cz+d}\right). \quad \text{[97][98]}$$

The complementary series are irreducible and inequivalent. As a representation of K, each is isomorphic to the Hilbert space direct sum of all the odd dimensional irreducible representations of $K = SU(2)$. Irreducibility can be proved by analyzing the action of $\mathfrak{g}$ on the algebraic sum of these subspaces[94][95] or directly without using the Lie algebra.[99][100]

11.2.7 Plancherel theorem

The only irreducible unitary representations of SL(2, **C**) are the principal series, the complementary series and the trivial representation. Since $-I$ acts $(-1)^k$ on the principal series and trivially on the remainder, these will give all the irreducible unitary representations of the Lorentz group, provided k is taken to be even.

To decompose the left regular representation of G on $L^2(G)$, only the principal series are required. This immediately yields the decomposition on the subrepresentations $L^2(G/\pm I)$, the left regular representation of the Lorentz group, and $L^2(G/K)$, the regular representation on 3-dimensional hyperbolic space. (The former only involves principal series representations with k even and the latter only those with $k = 0$.)

The left and right regular representation λ and ρ are defined on $L^2(G)$ by

$$\lambda(g)f(x) = f(g^{-1}x), \ \rho(g)f(x) = f(xg).$$

Now if f is an element of $C_c(G)$, the operator $\pi v,k(f)$ defined by

$$\pi_{\nu,k}(f) = \int_G f(g)\pi(g)\,dg$$

is Hilbert–Schmidt. We define a Hilbert space H by

$$H = \bigoplus_{k \geq 0} HS(L^2(C)) \otimes L^2(R, c_k(\nu^2 + k^2)^{1/2}d\nu),$$

where

$$c_0 = 1/4\pi^{3/2}, \ c_k = 1/(2\pi)^{3/2} \ (k \neq 0)$$

and HS(L^2(**C**)) denotes the Hilbert space of Hilbert–Schmidt operators on L^2(**C**).[nb 21] Then the map U defined on $C_c(G)$ by

$$U(f)(\nu, k) = \pi_{\nu,k}(f)$$

extends to a unitary of $L^2(G)$ onto H.

The map U satisfies

$$U(\lambda(x)\rho(y)f)(\nu, k) = \pi_{\nu,k}(x)^{-1}\pi_{\nu,k}(f)\pi_{\nu,k}(y).$$

If f_1, f_2 are in $C_c(G)$ then

$$(f_1, f_2) = \sum_{k \geq 0} c_k^2 \int_{-\infty}^{\infty} \mathrm{Tr}(\pi_{\nu,k}(f_1)\pi_{\nu,k}(f_2)^*)(\nu^2 + k^2)\,d\nu.$$

Thus if $f = f_1 * f_2*$ denotes the convolution of f_1 and f_2*, and $f_2^*(g) = \overline{f_2(g^{-1})}$, then

$$f(1) = \sum_{k \geq 0} c_k^2 \int_{-\infty}^{\infty} \mathrm{Tr}(\pi_{\nu,k}(f))(\nu^2 + k^2)\,d\nu.$$

The last two displayed formulas are usually referred to as the Plancherel formula and the Fourier inversion formula respectively. The Plancherel formula extends to all fi in $L_2(G)$. By a theorem of Jacques Dixmier and Paul Malliavin, every function f in $C_c^\infty(G)$ is a finite sum of convolutions of similar functions, the inversion formula holds for such f. It can be extended to much wider classes of functions satisfying mild differentiability conditions.[27]

11.3   Explicit formulas

11.3.1   Conventions and Lie algebra bases

The metric of choice is given by $\eta = \mathrm{diag}(-1, 1, 1, 1)$, and the physics convention for Lie algebras and the exponential mapping is used in this article. These choices are arbitrary, but once they are made, fixed. The rationale is to allow the use of a single reference[28] for several related formulas. One possible choice of basis for the Lie algebra (which is not fixed by the reference) is, in the 4-vector representation, given by

$$J_1 = J^{23} = -J^{32} = i\begin{pmatrix} 0&0&0&0\\0&0&0&0\\0&0&0&-1\\0&0&1&0 \end{pmatrix},$$

$$J_2 = J^{31} = -J^{13} = i\begin{pmatrix} 0&0&0&0\\0&0&0&1\\0&0&0&0\\0&-1&0&0 \end{pmatrix},$$

$$J_3 = J^{12} = -J^{21} = i\begin{pmatrix} 0&0&0&0\\0&0&-1&0\\0&1&0&0\\0&0&0&0 \end{pmatrix},$$

$$K_1 = J^{01} = J^{10} = i\begin{pmatrix} 0&1&0&0\\1&0&0&0\\0&0&0&0\\0&0&0&0 \end{pmatrix},$$

$$K_2 = J^{02} = J^{20} = i\begin{pmatrix} 0&0&1&0\\0&0&0&0\\1&0&0&0\\0&0&0&0 \end{pmatrix},$$

$$K_3 = J^{03} = J^{30} = i\begin{pmatrix} 0&0&0&1\\0&0&0&0\\0&0&0&0\\1&0&0&0 \end{pmatrix}.$$

The commutation relations of the Lie algebra $\mathbf{so}(3;1)$ are

$$[J^{\mu\nu}, J^{\rho\sigma}] = i(\eta^{\sigma\mu}J^{\rho\nu} + \eta^{\nu\sigma}J^{\mu\rho} - \eta^{\rho\mu}J^{\sigma\nu} - \eta^{\nu\rho}J^{\mu\sigma}). \text{ [28]}$$

In three-dimensional notation, these are

$$[J_i, J_j] = i\epsilon_{ijk}J_k, \quad [J_i, K_j] = i\epsilon_{ijk}K_k, \quad [K_i, K_j] = -i\epsilon_{ijk}J_k. \text{ [28]}$$

The choice of basis above satisfies the relations, but other choices are possible. The multiple use of the symbol J above and in the sequel should be observed.

Let $\pi_{(m, n)} : \mathbf{so}(3;1) \to \mathbf{gl}(V)$, where V is a vector space, denote the irreducible representations of $\mathbf{so}(3;1)$ according to the (m, n) classification. In components, with $-m \le a,a' \le m$, $-n \le b,b' \le n$, the representations are given by

$$(\pi_{m,n}(J_i))_{a'b',ab} = \delta_{b'b}(J_i^{(m)})_{a'a} + \delta_{a'a}(J_i^{(n)})_{b'b}, \quad \text{[28]}$$
$$(\pi_{m,n}(K_i))_{a'b',ab} = i(\delta_{a'a}(J_i^{(n)})_{b'b} - \delta_{b'b}(J_i^{(m)})_{a'a}),$$

where δ is the Kronecker delta and the $Ji^{(n)}$ are the $(2n + 1)$-dimensional irreducible representations of $\mathbf{so}(3)$, also termed **spin matrices** or **angular momentum matrices**. These are explicitly given by

$$(J_3^{(j)})_{a'a} = a\delta_{a'a},$$
$$(J_1^{(j)} \pm iJ_2^{(j)})_{a'a} = \sqrt{(j \mp a)(j \pm a + 1)}\delta_{a',a\pm1}. \quad \text{[28]}$$

11.3.2   Weyl spinors and bispinors

By taking, in turn, $m = 1/2$, $n = 0$ and $m = 0$, $n = 1/2$ and by setting

$$J_i^{(\frac{1}{2})} = \frac{1}{2}\sigma_i$$

in the general expression **(G1)**, and by using the trivial relations $1_1 = 1$ and $J^{(0)} = 0$, one obtains

$$\pi_{(\frac{1}{2},0)}(J_i) = \frac{1}{2}(\sigma_i \otimes 1_{(1)} + 1_{(2)} \otimes J_i^{(0)}) = \frac{1}{2}\sigma_i \quad \pi_{(\frac{1}{2},0)}(K_i) = \frac{i}{2}(1_{(2)} \otimes J_i^{(0)} - \sigma_i \otimes 1_{(1)}) = -\frac{i}{2}\sigma_i,$$

$$\pi_{(0,\frac{1}{2})}(J_i) = \frac{1}{2}(J_i^{(0)} \otimes 1_{(2)} + 1_{(1)} \otimes \sigma_i) = \frac{1}{2}\sigma_i \quad \pi_{(0,\frac{1}{2})}(K_i) = \frac{i}{2}(1_{(1)} \otimes \sigma_i - J_i^{(0)} \otimes 1_{(2)}) = +\frac{i}{2}\sigma_i.$$

These are the left-handed and right-handed Weyl spinor representations. They act by matrix multiplication on 2-dimensional complex vector spaces (with a choice of basis) VL and VR, whose elements ΨL and ΨR are called left- and right-handed Weyl spinors respectively. Given $(\pi(1/2,0), VL)$ and $(\pi(0,1/2), VR)$ one may form their direct sum as representations,

$$\pi_{(\frac{1}{2},0)\oplus(0,\frac{1}{2})}(J_i) = \frac{1}{2}\begin{pmatrix} \sigma_i & 0 \\ 0 & \sigma_i \end{pmatrix}.$$

$$\pi_{(\frac{1}{2},0)\oplus(0,\frac{1}{2})}(K_i) = \frac{i}{2}\begin{pmatrix} \sigma_i & 0 \\ 0 & -\sigma_i \end{pmatrix}$$

This is, up to a similarity transformation, the $(1/2,0) \oplus (0,1/2)$ Dirac spinor representation of **so**(3; 1). It acts on the 4-component elements $(\Psi L, \Psi R)$ of $(VL \oplus VR)$, called bispinors, by matrix multiplication. The representation may be obtained in a more general and basis independent way using Clifford algebras. These expressions for bispinors and Weyl spinors all extend by linearity of Lie algebras and representations to all of **so**(3; 1). Expressions for the group representations are obtained by exponentiation.

11.4 See also

- Bargmann–Wigner equations

- Center of mass (relativistic)

- Dirac algebra

- Gamma matrices

- Lorentz group

- Möbius transformation

- Poincaré group

- Representation theory of the Poincaré group

- Symmetry in quantum mechanics

- Wigner's classification

11.5 Remarks

[1] Tensor products of representations, $\pi g \otimes \pi h$ of $\mathbf{g} \oplus \mathbf{h}$ can, when both factors come from the same Lie algebra ($\mathbf{h} = \mathbf{g}$), either be thought of as a representation of $\mathbf{g}$ or $\mathbf{g} \oplus \mathbf{g}$.

[2] The "traceless" property can be expressed as $S\alpha\beta g^{\alpha\beta} = 0$, or $S\alpha^\alpha = 0$, or $S^{\alpha\beta}g\alpha\beta = 0$ depending on the presentation of the field: covariant, mixed, and contravariant respectively.

[3] This is provided parity is a symmetry. Else there would be two flavors, (3/2, 0) and (0, 3/2) in analogy with neutrinos.

[4] It's a rather deep fact that all finite-dimensional Lie algebras are linear. See Ado's theorem. The corresponding statement for compact Lie groups is true, but not for general Lie groups.

[5] Hall 2003, Equation 2.16. Due to the physicist conventions, the formula here differs with a factor of i in the exponent.

[6] In particular, A commutes with the Pauli matrices, hence with all of SU(2) making Schur's lemma applicable.

[7] The kernel of a Lie algebra homomorphism is an ideal, hence a subspace. Since **p** is 2:1 and both SL(2, **C**) and SO(3; 1)$^+$ are 6-dimensional, the kernel must be 0-dimensional, hence {∅}.

[8] The exponential map is one-to-one in a neighborhood if the identity in SL(2, **C**), hence the composition exp ∘ σ ∘ log:SL(2, **C**) → SO(3; 1)$^+$, where σ is the Lie algebra isomorphism, is onto an open neighborhood $U \subset$ SO(3; 1)$^+$ containing the identity. Such a neighborhood generates the connected component.

[9] Rossmann 2002 From Section 2.1 Example 4: This can be seen as follows. The matrix q has eigenvalues $\{-1, -1\}$, but it is not diagonalizable. If $q = \exp(Q)$, then Q has eigenvalues λ, $-\lambda$ with $\lambda = i\pi + 2\pi ik$ for some k because the tracelessness of **sl**(2, **C**)-matrices forces them to be negatives of each other. But then Q is diagonalizable, hence q is diagonalizable. This is a contradiction.

[10] Rossmann 2002 Proposition 10, paragraph 6.3. This is a consequence of the Peter-Weyl theorem.

[11] Hall 2003 Any discrete normal subgroup of a path connected group G is contained in the center Z of G. See Exercise 11, Chapter 1.

[12] A semisimple Lie group does not have any non-discrete normal abelian subgroups. This can be taken as the definition of semisimplicity.

[13] A simple group does not have any non-discrete normal subgroups.

[14] Lee 2003 Lemma A.17 (c). Closed subsets of compact sets are compact.

[15] Lee 2003 Lemma A.17 (a). If $f:X \to Y$ is continuous, X is compact, then $f(X)$ is compact.

[16] This is one of the conclusions of Cartan's theorem, the theorem of the highest weight. See Hall (2003) Chapter 6.

[17] Hall 2003 The root system is the union of two copies of A_1, where each copy resides in its own dimensions in the embedding vector space.

[18] Rossmann 2003 This definition is equivalent to the definition in terms of the connected Lie group whose Lie algebra is the Lie algebra of the root system under consideration.

[19] The elements of $L^2(\mathbf{S}^2)$ are actually equivalence classes of functions. two functions are declared equivalent if they differ merely on a set of measure zero. The integral is the Lebesgue integral in order to obtain a *complete* inner product space.

[20] A Hilbert space is separable if and only if it has a countable basis. All separable Hilbert spaces are isomorphic.

[21] Note that for a Hilbert space H, HS(H) may be identified canonically with the Hilbert space tensor product of H and its conjugate space.

11.6   Notes

[1] Greiner 1996

[2] These facts can be found in most elementary mathematics texts, and many physics texts. See e.g. Rossmann (2002), Hall (2003)

[3] See e.g. Hall (2003)

[4] Wigner 1937

[5] Hall 2003 Appendix D2.

[6] Weinberg 2003

[7] Lie 1888, 1890, 1893

[8] Killing 1888

[9] Cartan 1913

[10] Brauer & Weyl 1935 *Spinors in n dimensions.*

[11] Weyl 1931 *The Theory of Groups and Quantum Mechanics.*

[12] Weyl 1939 *The Classical Groups. Their Invariants and Representations.*

[13] Harish-Chandra 1947 *Infinite irreducible representations of the Lorentz group.*

[14] Wigner 1939 *On unitary representations of the inhomogeneous Lorentz group.*

[15] Bargmann 1947 *Irreducible unitary representations of the Lorenz group.*

[16] Bargmann was also a mathematician. He worked as Albert Einsteins assistant at the Institute for Advanced Study in Princeton.

[17] Bargmann & Wigner 1948 *Group theoretical discussion of relativistic wave equations.*

[18] Dirac 1928

[19] Weinberg 2003 Equation 5.6.7-8

[20] Weinberg 2003, Chapter 5

[21] Weinberg 2003 Equation 5.6.9-11

[22] Hall 2003, Chapter 6

[23] Knapp 2001 The rather mysterious looking third isomorphism is proved in chapter 2, paragraph 4.

[24] Hall 2003, Chapter 4

[25] Rossmann 2002 Section 6.5

[26] Knapp 2001 See section 2.3

[27] Knapp 2001

[28] Weinberg 2002, Chapter 5

[29] Weinberg 2003 See footnote on p. 232.

[30] Lie 1888

[31] Rossmann 2002, Section 2.5

[32] Rossmann 2002 Theorem 1, Paragraph 2.5.

[33] Rossmann 2002 Proposition 3, Paragraph 2.5.

[34] Rossmann 2002 Theorem 1, Paragraph 2.6.

[35] Hall 2003, Theorem 2.27, Chapter 2.

[36] Hall 2003 Section 3.1.

[37] Hall 2003 Formulae 3.1, 3.2 and 3.3 modified for physics convention.

[38] Hall 2003 See section 3.6 for a detailed account.

[39] Hall 2003 Exercise 10 Chapter 3.

[40] Rossmann 2002

[41] Hall 2003 Step 5 in proof of theorem 3.7

[42] Weinberg 2002, Appendix B, Chapter 2

[43] Weinberg 2002, Section 2.7, Chapter 2

[44] Wigner 1937, p.27

[45] Weinberg 2003, Section 2.7

[46] Gelfand, Minlos & Shapiro 1963 This construction of the covering group is treated in paragraph 4, section 1, chapter 1 in Part II.

[47] Hall 2003 Chapter 4.

[48] Knapp 2001 Chapter 2.

[49] Knapp 2001 Equation 2.1.

[50] Hall 2003 Equation 4.2.

[51] Hall 2003 Equation after 4.2.

[52] Hall 2003 Equation 4.3.

[53] Hall 2003 Equation after 4.3

[54] Knapp 2001 Equation 2.4.

[55] Rossmann 2002 Section 2.1

[56] Rossmann 2002 Section 2.1 Example 4.

[57] Hall 2003 Corollary 2.31.

[58] Hall 2003 Appendix A.

[59] Weinberg 2003 Chapter 5.

[60] Hall 2003 Theorem 7.15.

[61] Hall 2003 Theorem 7.43.

[62] Hall 2003 p. 235.

[63] Rossmann 2002 Propositions 3 and 6 paragraph 2.5.

[64] Hall 2003 See exercise 1, Chapter 6.

[65] Wigner 1937

[66] Bekaert & Boulanger 2006 p.4.

[67] Hall 2003 Proposition 1.20.

[68] Lee 2003 Theorem 8.30

[69] Weinberg 2002Chapters 2, 5.

[70] Weinberg 2002 This is outlined (very briefly) on page 232, hardly more than a footnote.

[71] Hall 2003 Proposition 7.39.

[72] Hall 2003 Theorem 7.40.

[73] Hall 2003 Theorem 7.40.

[74] Hall 2003 See section 6.6 for a detailed discussion.

[75] Hall 2003 Second item in proposition 4.5.

[76] Hall 2003 p.219

[77] Rossmann 2002 See exercise 3 in paragraph 6.5.

[78] Hall 2003 See appendix D.3

[79] Weinberg 2002 Section 5.4.

[80] Weinberg 2002 Eqn 5.4.6

[81] Weinberg 2003 Equation 2.6.5.

[82] Weinberg 2003 Equation following 2.6.6.

[83] Weinberg 2002 Chapter 2.

[84] Greiner, W (2000), *Relativistic Quantum Mechanics*, ISBN 3-540-67457-8

[85] For a detailed discussion of the spin 0, 1/2 and 1 cases, see Greiner & Reinhardt 1996.

[86] Weinberg 2002, Chapter 3

[87] Rossmann 2002 See section 6.1 for more examples, both finite-dimensional and infinite-dimensional.

[88] Gelfand, Minlos & Shapiro 1963

[89] In *Quantum Mechanics - non-relativistic theory* by Landau and Lifshitz the lowest order D are calculated analytically.

[90] Curtright, Fairlie & Zachos 2014 A formula for $D^{(\ell)}$ valid for all ℓ is given.

[91] Hall 2003 Section 4.3.5.

[92] Gelfand & Graev 1969

[93] Knapp 2001, Chapter II

[94] Harish-Chandra 1947

[95] Taylor 1986

[96] Knapp 2001 Chapter 2. Equation 2.12.

[97] Bargmann 1947

[98] Gelfand & Gaev 1963

[99] Gelfand & Naimark 1947

[100] Takahashi 1963, p. 343

[101] Weinberg 2002, Equations (5.4.19) and (5.4.20)

11.7   Freely available online references

- Bekaert, X.; Boulanger, N. (2006). "The unitary representations of the Poincare group in any spacetime dimension". arXiv:hep-th/0611263. Expanded version of the lectures presented at the second Modave summer school in mathematical physics (Belgium, August 2006).

- Curtright, T L; Fairlie, D B; Zachos, C K (2014), "A compact formula for rotations as spin matrix polynomials", *SIGMA* **10**: 084, arXiv:1402.3541, Bibcode:2014SIGMA..10..084C, doi:10.3842/SIGMA.2014.084 Group elements of SU(2) are expressed in closed form as finite polynomials of the Lie algebra generators, for all definite spin representations of the rotation group.

11.8   References

- Bargmann, V. (1947), "Irreducible unitary representations of the Lorenz group", *Ann. Of Math.* **48** (3): 568–640, doi:10.2307/1969129, JSTOR 1969129 (the representation theory of SO(2,1) and SL(2, **R**); the second part on SO(3; 1) and SL(2, **C**), described in the introduction, was never published).

- Bargmann, V.; Wigner, E. P. (1948), "Group theoretical discussion of relativistic wave equations", *Proc. Natl. Sci. U. S. A.* **34** (5): 211–23

- Brauer, R.; Weyl, H. (1935), "Spinors in n dimensions", *Amer. J. Math.* **57** (2): 425–449, doi:10.2307/2371218

- Cartan, Élie (1913), "Les groupes projectifs qui ne laissant invariante aucun multiplicité plane", *Bull. Soc. Math.* (in French) **41**: 53–96

- Dirac, P. A. M. (1928), "The Quantum Theory of the Electron", *Proc. Roy. Soc. A* **117** (778): doi:10.1098/rspa.1928.0023 (free access)

- Dirac, P. A. M. (1936), *Relativistic wave equations* **155** (886), Proc. Roy. Soc. A, pp. 447–459, doi:10.1098/rspa.1936.0111

- Dirac, P. A. M. (1945), *Unitary representations of the Lorentz group* **183** (994), Proc. Roy. Soc. A, pp. 284–295, Bibcode:1945RSPSA.183..284D, doi:10.1098/rspa.1945.0003

- Dixmier, J.; Malliavin, P. (1978), "Factorisations de fonctions et de vecteurs indéfiniment différentiables", *Bull. Sc. Math.* (in French) **102**: 305–330

- Fierz, M. (1939), "Über die relativistische theorie Kräftefreier teilchen mit beliebigem spin", *Helv. Phys. Acta.* (in German) **12** (1): 3–37, doi:10.5169/seals-110930(pdf download available)

- Fierz, M.; Pauli, W. (1939), "On relativistic wave equations for particles of arbitrary spin in an electromagnetic field", *Proc.Roy. Soc. A* **173** (953): 211–232, Bibcode:1939RSPSA.173..211F, doi:10.1098/rspa.1939.0140

- Gelfand, I. M.; M. I. Graev (1953), "On a general method of decomposition of the regular representation of a Lie group into irreducible representations", *Doklady Akademii Nauk SSSR* **92**: 221–224

- Gelfand, I. M.; Graev, M. I.; Pyatetskii-Shapiro, I. I. (1969), *Representation theory and automorphic functions*, Academic Press, ISBN 0-12-279506-7

- Gelfand, I.M.; Minlos, R.A.; Shapiro, Z. Ya. (1963), *Representations of the Rotation and Lorentz Groups and their Applications*, New York: Pergamon Press

- Gelfand, I. M.; Naimark, M. A. (1947), "Unitary representations of the Lorentz group" (PDF), *Izvestiya Akad. Nauk SSSR. Ser. Mat.* (in Russian) **11** (5): 411–504, retrieved 2014-12-15(Pdf from Math.net.ru)

- Greiner, W.; Reinhardt, J. (1996), *Field Quantization*, Springer, ISBN 3-540-59179-6

- Harish-Chandra (1947), "Infinite irreducible representations of the Lorentz group", *Proc. Roy. Soc. A* **189** (1018): 372–401, Bibcode:1947RSPSA.189..372H, doi:10.1098/rspa.1947.0047

- Harish-Chandra (1951), "Plancherel formula for complex semi-simple Lie groups", *Proc. Nat. Acad. Sci. U. S. A.* **37** (12): 813–818, Bibcode:1951PNAS...37..813H, doi:10.1073/pnas.37.12.813

- Hall, Brian C. (2003), *Lie Groups, Lie Algebras, and Representations An Elementary Introduction*, Springer, ISBN 0-387-40122-9

- Helgason, S. (1968), *Lie groups and symmetric spaces*, Battelle Rencontres, Benjamin, pp. 1–71 (a general introduction for physicists)

- Helgason, S. (2000), *Groups and geometric analysis. Integral geometry, invariant differential operators, and spherical functions (corrected reprint of the 1984 original)*, Mathematical Surveys and Monographs **83**, American Mathematical Society, ISBN 0-8218-2673-5

- Jorgenson, J.; Lang, S. (2008), *The heat kernel and theta inversion on SL(2,C)*, Springer Monographs in Mathematics, Springer, ISBN 978-0-387-38031-5

- Killing, Wilhelm (1888), "Die Zusammensetzung der stetigen/endlichen Transformationsgruppen", *Mathematische Annalen* (in German) **31** (2 (June)): 252–290, doi:10.1007/bf01211904

- Knapp, Anthony W. (2001), *Representation theory of semisimple groups. An overview based on examples.*, Princeton Landmarks in Mathematics, Princeton University Press, ISBN 0-691-09089-0 (elementary treatment for SL(2,**C**))

- Lee, J. M. (2003), *Introduction to Smooth manifolds*, Springer Graduate Texts in Mathematics **218**, ISBN 0-387-95448-1

- Lie, Sophus (1888), *Theorie der Transformationsgruppen I(1888), II(1890), III(1893)* (in German)

- Naimark, M.A. (1964), *Linear representations of the Lorentz group (translated from the Russian original by Ann Swinfen and O. J. Marstrand)*, Macmillan

- Rossmann, Wulf (2002), *Lie Groups - An Introduction Through Linear Groups*, Oxford Graduate Texts in Mathematics, Oxford Science Publications, ISBN 0 19 859683 9

- Rühl, W. (1970), *The Lorentz group and harmonic analysis*, Benjamin (a detailed account for physicists)

- Takahashi, R. (1963), "Sur les représentations unitaires des groupes de Lorentz généralisés", *Bull. Soc. Math. France* (in French) **91**: 289–433

- Taylor, M. E. (1986), *Noncommutative harmonic analysis*, Mathematical Surveys and Monographs **22**, American Mathematical Society, ISBN 0-8218-1523-7, Chapter 9, SL(2, **C**) and more general Lorentz groups

- Weinberg, S. (2002), *The Quantum Theory of Fields* **1**, Cambridge University Press, ISBN 0-521-55001-7

- Weyl, H. (1939), *The Classical Groups. Their Invariants and Representations*, Princeton University Press, ISBN 978-0-691-05756-9, MR 0000255

- Weyl, H. (1931), *The Theory of Groups and Quantum Mechanics*, Dover, ISBN 0-486-60269-9

- Wigner, E. P. (1939), "On unitary representations of the inhomogeneous Lorentz group", *Annals of Mathematics* **40** (1): 149 204, Bibcode:1939AnMat..40..922E, doi:10.2307/1968551, MR 1503456.

E.P. Wigner investigated the Lorentz group in depth and is known for the Bargmann-Wigner equations. The realization of the covering group given here is from his 1937 paper.

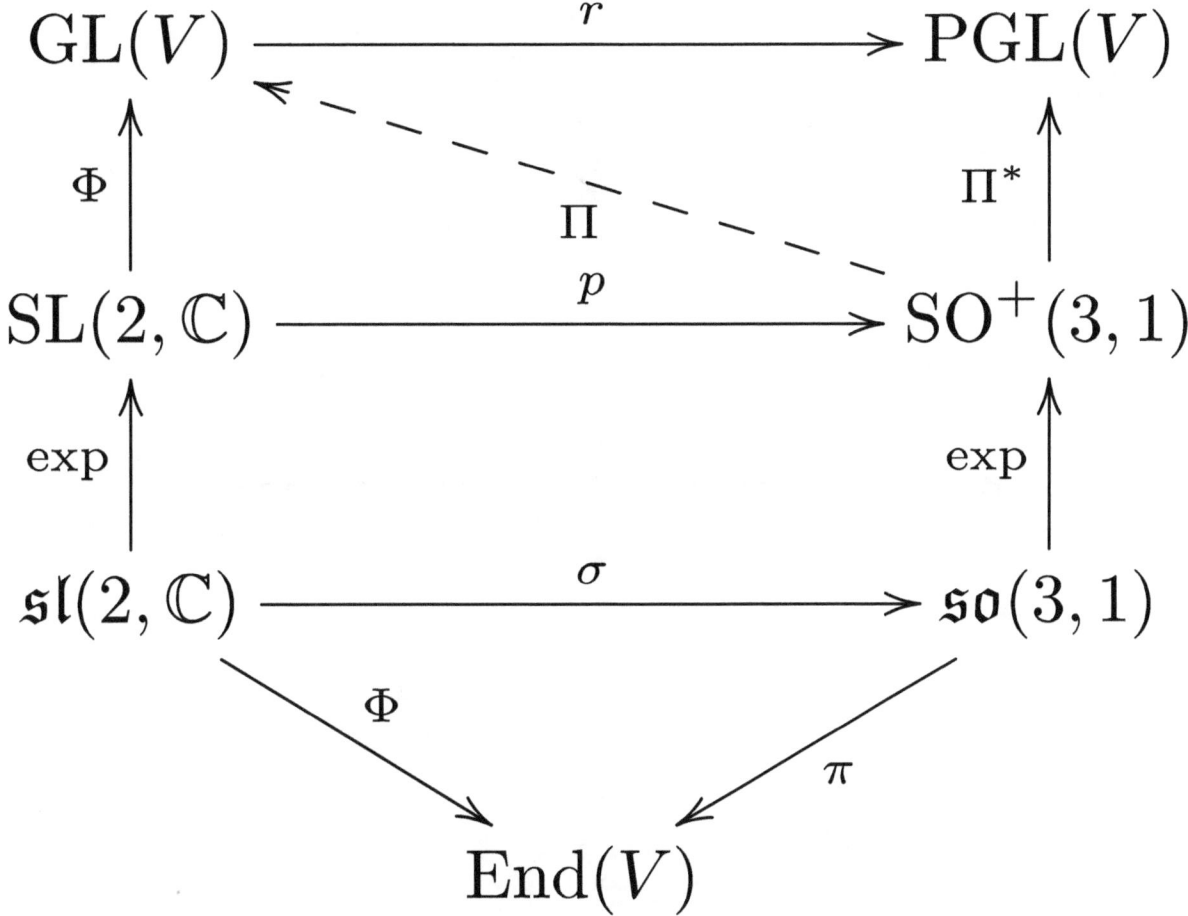

This diagram shows the web of maps discussed in the text. Here V is a finite-dimensional vector space carrying representations of sl(2, C), so(3; 1), SL(2, C), SO(3; 1)⁺, exp is the exponential mapping, p is the covering map from SL(2, C) onto SO(3; 1)⁺ and σ is the Lie algebra isomorphism induced by it. The maps Π, π and the two Φ are representations. the picture is only partially true when Π is projective.

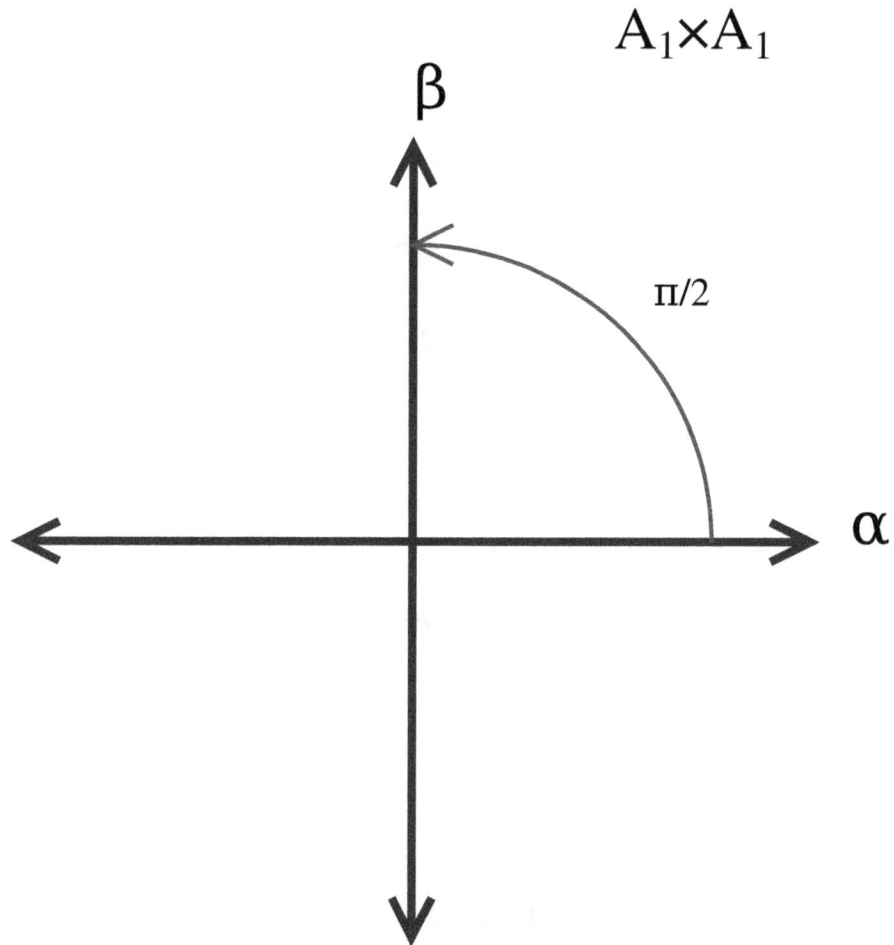

*The root system A1 × A1 of sl(2, C) ⊕ **sl(2, C)**.*

Richard Brauer and wife Ilse 1970. Brauer generalized the spin representations of Lie algebras sitting inside Clifford algebras to spin higher than 1/2. Photo courtesy of MFO.

Israïl Moiseevich Gelfand and Mark Naimark found the Plancherel formula for the Lorentz group in 1947.

P.A.M Dirac put the (1/2, 0) ⊕ (0, 1/2)-representation to use in the Dirac equation.

Chapter 12

Representation theory of the Poincaré group

In mathematics, **the representation theory of the Poincaré group** is an example of the representation theory of a Lie group that is neither a compact group nor a semisimple group. It is fundamental in theoretical physics.

In a physical theory having Minkowski space as the underlying spacetime, the space of physical states is typically a representation of the Poincaré group. (More generally, it may be a projective representation, which amounts to a representation of the double cover of the group.)

In a classical field theory, the physical states are sections of a Poincaré-equivariant vector bundle over Minkowski space. The equivariance condition means that the group acts on the total space of the vector bundle, and the projection to Minkowski space is an equivariant map. Therefore the Poincaré group also acts on the space of sections. Representations arising in this way (and their subquotients) are called covariant field representations, and are not usually unitary.

For a discussion of such unitary representations, see Wigner's classification.

In quantum mechanics, the state of the system is determined by the Schrödinger equation, which is only invariant under Galilean transformations. Quantum field theory is the relativistic extension of quantum mechanics, where relativistic (Lorentz/Poincaré invariant) wave equations are solved, "quantized", and act on a Hilbert space composed of Fock states; eigenstates of the theory's Hamiltonian which are states with a definite number of particles with individual 4-momentum. There are no finite unitary representations of the full Lorentz (and thus Poincaré) transformations due to the non-compact nature of Lorentz boosts (rotations in Minkowski space along a space and time axis).

In case of spin 1/2 particles, it is possible to find a construction that includes both a finite-dimensional representation and a scalar product preserved by this representation by associating a 4-component Dirac spinor ψ with each particle. These spinors transform under Lorentz transformations generated by the gamma matrices (γ_μ). It can be shown that the scalar product

$$\langle \psi | \phi \rangle = \bar{\psi}\phi = \psi^\dagger \gamma_0 \phi$$

is preserved. It is not, however, positive definite, so the representation is not unitary.

12.1   See also

- Wigner's classification

- Representation theory of the Lorentz group

- Representation theory of the Galilean group

- Representation theory of diffeomorphism groups

- Particle physics and representation theory

- Symmetry in quantum mechanics

- Center of mass (relativistic)

Chapter 13

Representation theory of the Galilean group

In nonrelativistic quantum mechanics, an account can be given of the existence of mass and spin (normally explained in Wigner's classification of relativistic mechanics) in terms of the **representation theory of the Galilean group**, which is the spacetime symmetry group of nonrelativistic quantum mechanics.

In 3+1 dimensions, this is the subgroup of the affine group on (t,x,y,z), whose linear part leaves invariant both the metric ($g\mu\nu=$ diag(1,0,0,0)) and the (independent) dual metric ($g\mu\nu=$ diag(0,1,1,1)). A similar definition applies for n+1 dimensions.

We are interested in projective representations of this group, which are equivalent to unitary representations of the nontrivial central extension of the universal covering group of the Galilean group by the one-dimensional Lie group **R**, cf. the article Galilean group for the central extension of its Lie algebra. The method of induced representations will be used to survey these.

We focus on the Lie algebra here because it is simpler to analyze and we can always extend the results to the full Lie group through the Frobenius theorem.

$$[E, P_i] = 0$$

$$[P_i, P_j] = 0$$

$$[L_{ij}, E] = 0$$

$$[C_i, C_j] = 0$$

$$[L_{ij}, L_{kl}] = i\hbar[\delta_{ik}L_{jl} - \delta_{il}L_{jk} - \delta_{jk}L_{il} + \delta_{jl}L_{ik}]$$

$$[L_{ij}, P_k] = i\hbar[\delta_{ik}P_j - \delta_{jk}P_i]$$

$$[L_{ij}, C_k] = i\hbar[\delta_{ik}C_j - \delta_{jk}C_i]$$

$$[C_i, E] = i\hbar P_i$$

$$[C_i, P_j] = i\hbar M\delta_{ij} \ .$$

E is the generator of time translations (Hamiltonian), *Pi* is the generator of translations (momentum operator), *Ci* is the generator of Galileian boosts, and *Lij* stands for a generator of rotations (angular momentum operator).

The central charge M is a Casimir invariant.

The mass-shell invariant

$$ME - \frac{P^2}{2}$$

is an additional Casimir invariant.

In 3+1 dimensions, a third Casimir invariant is W^2, where

$$\vec{W} \equiv M\vec{L} + \vec{P} \times \vec{C},$$

somewhat analogous to the Pauli–Lubanski pseudovector of relativistic mechanics.

More generally, in n+1 dimensions, invariants will be a function of

$$W_{ij} = ML_{ij} + P_i C_j - P_j C_i$$

and

$$W_{ijk} = P_i L_{jk} + P_j L_{ki} + P_k L_{ij},$$

as well as of the above mass-shell invariant and central charge.

Using Schur's lemma, in an irreducible unitary representation, all these Casimir invariants are multiples of the identity. Call these coefficients m and mE_0 and (in the case of 3+1 dimensions) w, respectively. Recalling that we are considering unitary representations here, we see that these eigenvalues have to be real numbers.

Thus, m > 0, m = 0 and m < 0. (The last case is similar to the first.) In 3+1 dimensions, when m >0, we can write, $w = ms$ for the third invariant, where s represents the spin, or intrinsic angular momentum. More generally, in n+1 dimensions, the generators L and C will be related, respectively, to the total angular momentum and center-of-mass moment by

$$W_{ij} = MS_{ij}$$
$$L_{ij} = S_{ij} + X_i P_j - X_j P_i$$
$$C_i = MX_i - P_i t.$$

From a purely representation-theoretic point of view, one would have to study all of the representations; but, here, we are only interested in applications to quantum mechanics. There, E represents the energy, which has to be bounded below, if thermodynamic stability is required. Consider first the case where m is nonzero.

Considering the (E, $P\rightarrow$) space with the constraint

$$mE = mE_0 + \frac{P^2}{2},$$

we see that the Galilean boosts act transitively on this hypersurface. In fact, treating the energy E as the Hamiltonian, differentiating with respect to P, and applying Hamilton's equations, we obtain the mass-velocity relation $mv\rightarrow = P\rightarrow$.

The hypersurface is parametrized by this velocity $v\rightarrow$. Consider the stabilizer of a point on the orbit, $(E_0, 0)$, where the velocity is 0. Because of transitivity, we know the unitary irrep contains a nontrivial linear subspace with these energy-momentum eigenvalues. (This subspace only exists in a rigged Hilbert space, because the momentum spectrum is continuous.)

The subspace is spanned by E, $P\rightarrow$, M and Lij. We already know how the subspace of the irrep transforms under all operators but the angular momentum. Note that the rotation subgroup is Spin(3). We have to look at its double cover, because we are considering projective representations. This is called the little group, a name given by Eugene Wigner. His method of induced representations specifies that the irrep is given by the direct sum of all the fibers in a vector bundle over the $mE = mE_0 + P^2/2$ hypersurface, whose fibers are a unitary irrep of Spin(3).

Spin(3) is none other than SU(2). (See representation theory of SU(2), where it is shown that the unitary irreps of SU(2) are labeled by s, a non-negative integer multiple of one half. This is called spin, for historical reasons.)

- Consequently, for m≠0, the unitary irreps are classified by m, E_0 and a spin s.

- Looking at the spectrum of E, it is evident that if m is negative, the spectrum of E is not bounded below. Hence, only the case with a positive mass is physical.

- Now, consider the case m = 0. By unitarity,

$$mE - \frac{P^2}{2} = \frac{-P^2}{2}$$

is nonpositive. Suppose it is zero. Here, it is also the boosts as well as the rotations that constitute the little group. Any unitary irrep of this little group also gives rise to a projective irrep of the Galilean group. As far as we can tell, only the case which transforms trivially under the little group has any physical interpretation, and it corresponds to the no-particle state, the vacuum.

The case where the invariant is negative requires additional comment. This corresponds to the representation class for m = 0 and non-zero $P\rightarrow$. Extending the bradyon, luxon, tachyon classification from the representation theory of the Poincaré group to an analogous classification, here, one may term these states as *synchrons*. They represent an instantaneous transfer of non-zero momentum across a (possibly large) distance. Associated with them, by above, is a "time" operator

$$t = -\frac{\vec{P} \cdot \vec{C}}{P^2} \, ,$$

which may be identified with the time of transfer. These states are naturally interpreted as the carriers of instantaneous action-at-a-distance forces.

N.B. In the 3+1-dimensional Galilei group, the boost generator may be decomposed into

$$\vec{C} = \frac{\vec{W} \times \vec{P}}{P^2} - \vec{P}t \, ,$$

with $W\rightarrow$ playing a role analogous to helicity.

13.1   See also

- Representation theory of the Poincaré group

- Wigner's classification

- Pauli–Lubanski pseudovector

- Representation theory of the diffeomorphism group

- Rotation operator

13.2   References

- Bargmann, V. (1954). "On Unitary Ray Representations of Continuous Groups", *Annals of Mathematics*, Second Series, **59**, No. 1 (Jan., 1954), pp. 1-46

- Lévy-Leblond, Jean-Marc (1967), "Nonrelativistic Particles and Wave Equations", *Communications in Mathematical Physics* (Springer) **6** (4): 286–311, Bibcode:1967CMaPh...6..286L, doi:10.1007/bf01646020.

- Ballentine, Leslie E. (1998). *Quantum Mechanics, A Modern Development.* World Scientific Publishing Co Pte Ltd. ISBN 981-02-4105-4.

- Gilmore, Robert (2006). *Lie Groups, Lie Algebras, and Some of Their Applications* (Dover Books on Mathematics) ISBN 0486445291

Chapter 14

List of simple Lie groups

In mathematics, the simple Lie groups were first classified by Wilhelm Killing and later perfected by Élie Cartan. This classification is often referred to as Killing-Cartan classification.

The list of simple Lie groups can be used to read off the list of simple Lie algebras and Riemannian symmetric spaces. See also the table of Lie groups for a smaller list of groups that commonly occur in theoretical physics, and the Bianchi classification for groups of dimension at most 3.

14.1 Simple Lie groups

Unfortunately, there is no generally accepted definition of a simple Lie group. In particular, it is not defined as a Lie group that is simple as an abstract group. Authors differ on whether a simple Lie group has to be connected, or on whether it is allowed to have a non-trivial center, or on whether **R** is a simple Lie group.

The most common definition is that a Lie group is simple if it is connected, non-abelian, and every closed connected normal subgroup is either the identity or the whole group. In particular, simple groups are allowed to have a non-trivial center.

In this article the connected simple Lie groups with trivial center are listed. Once these are known, the ones with non-trivial center are easy to list as follows. Any simple Lie group with trivial center has a universal cover, whose center is the fundamental group of the simple Lie group. The corresponding simple Lie groups with non-trivial center can be obtained as quotients of this universal cover by a subgroup of the center.

14.2 Simple Lie algebras

The Lie algebra of a simple Lie group is a simple Lie algebra. This is a one-to-one correspondence between connected simple Lie groups with trivial center and simple Lie algebras of dimension greater than 1. (Authors differ on whether the one-dimensional Lie algebra should be counted as simple.)

Over the complex numbers the simple Lie algebras are classified by their Dynkin diagrams, of types "ABCDEFG". If L is a real simple Lie algebra, its complexification is a simple complex Lie algebra, unless L is already the complexification of a Lie algebra, in which case the complexification of L is a product of two copies of L. This reduces the problem of classifying the real simple Lie algebras to that of finding all the real forms of each complex simple Lie algebra (i.e., real Lie algebras whose complexification is the given complex Lie algebra). There are always at least 2 such forms: a split form and a compact form, and there are usually a few others. The different real forms correspond to the classes of automorphisms of order at most 2 of the complex Lie algebra.

14.3 Symmetric spaces

Symmetric spaces are classified as follows.

First, the universal cover of a symmetric space is still symmetric, so we can reduce to the case of simply connected symmetric spaces. (For example, the universal cover of a real projective plane is a sphere.)

Second, the product of symmetric spaces is symmetric, so we may as well just classify the irreducible simply connected ones (where irreducible means they cannot be written as a product of smaller symmetric spaces).

The irreducible simply connected symmetric spaces are the real line, and exactly two symmetric spaces corresponding to each *non-compact* simple Lie group G, one compact and one non-compact. The non-compact one is a cover of the quotient of G by a maximal compact subgroup H, and the compact one is a cover of the quotient of the compact form of G by the same subgroup H. This duality between compact and non-compact symmetric spaces is a generalization of the well known duality between spherical and hyperbolic geometry.

14.4 Hermitian symmetric spaces

A symmetric space with a compatible complex structure is called Hermitian. The compact simply connected irreducible Hermitian symmetric spaces fall into 4 infinite families with 2 exceptional ones left over, and each has a non-compact dual. In addition the complex plane is also a Hermitian symmetric space; this gives the complete list of irreducible Hermitian symmetric spaces.

The four families are the types A III, B I and D I for $p = 2$, D III, and C I, and the two exceptional ones are types E III and E VII of complex dimensions 16 and 27.

14.5 Notation

R, C, H, and O stand for the real numbers, complex numbers, quaternions, and octonions.

In the symbols such as $E_6{}^{-26}$ for the exceptional groups, the exponent -26 is the signature of an invariant symmetric bilinear form that is negative definite on the maximal compact subgroup. It is equal to the dimension of the group minus twice the dimension of a maximal compact subgroup.

The fundamental group listed in the table below is the fundamental group of the simple group with trivial center. Other simple groups with the same Lie algebra correspond to subgroups of this fundamental group (modulo the action of the outer automorphism group).

14.6 List

See also: Abelian group

14.6.1 Compact

See also: Compact group

14.6.2   Split

See also: Split Lie algebra

14.6.3   Complex

See also: Complex Lie group

14.6.4   Others

14.7   Simple Lie groups of small dimension

The following table lists some Lie groups with simple Lie algebras of small dimension. The groups on a given line all have the same Lie algebra. In the dimension 1 case, the groups are abelian and not simple.

14.8   Notes

^† The group **R** is not simple as an abstract group, and according to most (but not all) definitions this is not a simple Lie group. Most authors do not count its Lie algebra as a simple Lie algebra. It is listed here so that the list of irreducible simply connected symmetric spaces is complete. Note that **R** is the only such non-compact symmetric space without a compact dual (although it has a compact quotient S^1).

14.9   Further reading

- Besse, *Einstein manifolds* ISBN 0-387-15279-2
- Helgason, *Differential geometry, Lie groups, and symmetric spaces.* ISBN 0-8218-2848-7
- Fuchs and Schweigert, *Symmetries, Lie algebras, and representations: a graduate course for physicists.* Cambridge University Press, 2003. ISBN 0-521-54119-0

Chapter 15

Symmetric space

For other uses, see Symmetric space (disambiguation).

In differential geometry, representation theory and harmonic analysis, a **symmetric space** is a smooth manifold whose group of symmetries contains an inversion symmetry about every point. There are two ways to formulate the inversion symmetry: via Riemannian geometry or via Lie theory. The Lie-theoretic definition is more general and more algebraic.

In Riemannian geometry, the inversions are geodesic symmetries, and these are required to be isometries, leading to the notion of a **Riemannian symmetric space**. More generally, in Lie theory a symmetric space is a homogeneous space G/H for a Lie group G such that the stabilizer H of a point is an open subgroup of the fixed point set of an involution of G. This definition includes (globally) Riemannian symmetric spaces and pseudo-Riemannian symmetric spaces as special cases.

Riemannian symmetric spaces arise in a wide variety of situations in both mathematics and physics. They were first studied extensively and classified by Élie Cartan. More generally, classifications of irreducible and semisimple symmetric spaces have been given by Marcel Berger. They are important in representation theory and harmonic analysis as well as differential geometry.

15.1 Definition using geodesic symmetries

Let M be a connected Riemannian manifold and p a point of M. A map f defined on a neighborhood of p is said to be a **geodesic symmetry**, if it fixes the point p and reverses geodesics through that point, i.e. if γ is a geodesic and $\gamma(0) = p$ then $f(\gamma(t)) = \gamma(-t)$. It follows that the derivative of the map at p is minus the identity map on the tangent space of p. On a general Riemannian manifold, f need not be isometric, nor can it be extended, in general, from a neighbourhood of p to all of M.

M is said to be **locally Riemannian symmetric** if its geodesic symmetries are in fact isometric, and **(globally) Riemannian symmetric** if in addition its geodesic symmetries are defined on all of M.

15.1.1 Basic properties

The Cartan–Ambrose–Hicks theorem implies that M is locally Riemannian symmetric if and only if its curvature tensor is covariantly constant, and furthermore that any simply connected, complete locally Riemannian symmetric space is actually Riemannian symmetric.

Any Riemannian symmetric space M is complete and Riemannian homogeneous (meaning that the isometry group of M acts transitively on M). In fact, already the identity component of the isometry group acts transitively on M (because M is connected).

Locally Riemannian symmetric spaces that are not Riemannian symmetric may be constructed as quotients of Riemannian symmetric spaces by discrete groups of isometries with no fixed points, and as open subsets of (locally) Riemannian symmetric spaces.

15.1.2 Examples

Basic examples of Riemannian symmetric spaces are Euclidean space, spheres, projective spaces, and hyperbolic spaces, each with their standard Riemannian metrics. More examples are provided by compact, semi-simple Lie groups equipped with a bi-invariant Riemannian metric. An example of a non-Riemannian symmetric space is anti-de Sitter space.

Any compact Riemann surface of genus greater than 1 (with its usual metric of constant curvature −1) is a locally symmetric space but not a symmetric space.

15.2 General definition

Let G be a connected Lie group. Then a **symmetric space** for G is a homogeneous space G/H where the stabilizer H of a typical point is an open subgroup of the fixed point set of an involution σ in $Aut(G)$. Thus σ is an automorphism of G with $\sigma^2 = \mathrm{id}G$ and H is an open subgroup of the set

$$G^\sigma = \{g \in G : \sigma(g) = g\}.$$

Because H is open, it is a union of components of G^σ (including, of course, the identity component).

As an automorphism of G, σ fixes the identity element, and hence, by differentiating at the identity, it induces an automorphism of the Lie algebra $\mathfrak{g}$ of G, also denoted by σ, whose square is the identity. It follows that the eigenvalues of σ are ± 1. The $+1$ eigenspace is the Lie algebra $\mathfrak{h}$ of H (since this is the Lie algebra of G^σ), and the -1 eigenspace will be denoted $\mathfrak{m}$. Since σ is an automorphism of $\mathfrak{g}$, this gives a direct sum decomposition

$$\mathfrak{g} = \mathfrak{h} \oplus \mathfrak{m}$$

with

$$[\mathfrak{h}, \mathfrak{h}] \subset \mathfrak{h}, \quad [\mathfrak{h}, \mathfrak{m}] \subset \mathfrak{m}, \quad [\mathfrak{m}, \mathfrak{m}] \subset \mathfrak{h}.$$

The first condition is automatic for any homogeneous space: it just says the infinitesimal stabilizer $\mathfrak{h}$ is a Lie subalgebra of $\mathfrak{g}$. The second condition means that $\mathfrak{m}$ is an $\mathfrak{h}$ -invariant complement to $\mathfrak{h}$ in $\mathfrak{g}$. Thus any symmetric space is a reductive homogeneous space, but there are many reductive homogeneous spaces which are not symmetric spaces. The key feature of symmetric spaces is the third condition that $\mathfrak{m}$ brackets into $\mathfrak{h}$.

Conversely, given any Lie algebra $\mathfrak{g}$ with a direct sum decomposition satisfying these three conditions, the linear map σ, equal to the identity on $\mathfrak{h}$ and minus the identity on $\mathfrak{m}$, is an involutive automorphism.

15.3 Riemannian symmetric spaces are symmetric spaces

If M is a Riemannian symmetric space, the identity component G of the isometry group of M is a Lie group acting transitively on M (M is Riemannian homogeneous). Therefore, if we fix some point p of M, M is diffeomorphic to the quotient G/K, where K denotes the isotropy group of the action of G on M at p. By differentiating the action at p we obtain an isometric action of K on T_pM. This action is faithful (e.g., by a theorem of Kostant, any isometry in the identity component is determined by its 1-jet at any point) and so K is a subgroup of the orthogonal group of T_pM, hence compact. Moreover, if we denote by sp: M → M the geodesic symmetry of M at p, the map

$$\sigma : G \to G, h \mapsto s_p \circ h \circ s_p$$

is an involutive Lie group automorphism such that the isotropy group K is contained between the fixed point group of σ and its identity component (hence an open subgroup).

To summarize, M is a symmetric space G/K with a compact isotropy group K. Conversely, symmetric spaces with compact isotropy group are Riemannian symmetric spaces, although not necessarily in a unique way. To obtain a Riemannian symmetric space structure we need to fix a K-invariant inner product on the tangent space to G/K at the identity coset eK: such an inner product always exists by averaging, since K is compact, and by acting with G, we obtain a G-invariant Riemannian metric g on G/K.

To show that G/K is Riemannian symmetric, consider any point $p = hK$ (a coset of K, where $h \in G$) and define

$$s_p : M \to M, h'K \mapsto h\sigma(h^{-1}h')K$$

where σ is the involution of G fixing K. Then one can check that sp is an isometry with (clearly) $sp(p) = p$ and (by differentiating) dsp equal to minus the identity on TpM. Thus sp is a geodesic symmetry and, since p was arbitrary, M is a Riemannian symmetric space.

If one starts with a Riemannian symmetric space M, and then performs these two constructions in sequence, then the Riemannian symmetric space yielded is isometric to the original one. This shows that the "algebraic data" (G,K,σ,g) completely describe the structure of M.

15.4 Classification of Riemannian symmetric spaces

Main article: List of simple Lie groups

The algebraic description of Riemannian symmetric spaces enabled Élie Cartan to obtain a complete classification of them in 1926.

For a given Riemannian symmetric space M let (G,K,σ,g) be the algebraic data associated to it. To classify possibly isometry classes of M, first note that the universal cover of a Riemannian symmetric space is again Riemannian symmetric, and the covering map is described by dividing the connected isometry group G of the covering by a subgroup of its center. Therefore, we may suppose without loss of generality that M is simply connected. (This implies K is connected by the long exact sequence of a fibration, because G is connected by assumption.)

15.4.1 Classification scheme

A simply connected Riemannian symmetric space is said to be **irreducible** if it is not the product of two or more Riemannian symmetric spaces. It can then be shown that any simply connected Riemannian symmetric space is a Riemannian product of irreducible ones. Therefore, we may further restrict ourselves to classifying the irreducible, simply connected Riemannian symmetric spaces.

The next step is to show that any irreducible, simply connected Riemannian symmetric space M is of one of the following three types:

1. **Euclidean type**: M has vanishing curvature, and is therefore isometric to a Euclidean space.

2. **Compact type**: M has nonnegative (but not identically zero) sectional curvature.

3. **Non-compact type**: M has nonpositive (but not identically zero) sectional curvature.

A more refined invariant is the **rank**, which is the maximum dimension of a subspace of the tangent space (to any point) on which the curvature is identically zero. The rank is always at least one, with equality if the sectional curvature is positive or negative. If the curvature is positive, the space is of compact type, and if negative, it is of noncompact type. The spaces of Euclidean type have rank equal to their dimension and are isometric to a Euclidean space of that dimension. Therefore, it remains to classify the irreducible, simply connected Riemannian symmetric spaces of compact and non-compact type. In both cases there are two classes.

A. G is a (real) simple Lie group;

B. G is either the product of a compact simple Lie group with itself (compact type), or a complexification of such a Lie group (non-compact type).

The examples in class B are completely described by the classification of simple Lie groups. For compact type, M is a compact simply connected simple Lie group, G is $M \times M$ and K is the diagonal subgroup. For non-compact type, G is a simply connected complex simple Lie group and K is its maximal compact subgroup. In both cases, the rank is the rank of G.

The compact simply connected Lie groups are the universal covers of the classical Lie groups $SO(n)$, $SU(n)$, $Sp(n)$ and the five exceptional Lie groups E_6, E_7, E_8, F_4, G_2.

The examples of class A are completely described by the classification of noncompact simply connected real simple Lie groups. For non-compact type, G is such a group and K is its maximal compact subgroup. Each such example has a corresponding example of compact type, by considering a maximal compact subgroup of the complexification of G which contains K. More directly, the examples of compact type are classified by involutive automorphisms of compact simply connected simple Lie groups G (up to conjugation). Such involutions extend to involutions of the complexification of G, and these in turn classify non-compact real forms of G.

In both class A and class B there is thus a correspondence between symmetric spaces of compact type and non-compact type. This is known as duality for Riemannian symmetric spaces.

15.4.2   Classification result

Specializing to the Riemannian symmetric spaces of class A and compact type, Cartan found that there are the following seven infinite series and twelve exceptional Riemannian symmetric spaces G/K. They are here given in terms of G and K, together with a geometric interpretation, if readily available. The labelling of these spaces is the one given by Cartan.

15.4.3   As Grassmannians

A more modern classification (Huang & Leung 2011) uniformly classifies the Riemannian symmetric spaces, both compact and non-compact, via a Freudenthal magic square construction. The irreducible compact Riemannian symmetric spaces are, up to finite covers, either a compact simple Lie group, a Grassmannian, a Lagrangian Grassmannian, or a double Lagrangian Grassmannian of subspaces of $(\mathbf{A} \otimes \mathbf{B})^n$, for normed division algebras $\mathbf{A}$ and $\mathbf{B}$. A similar construction produces the irreducible non-compact Riemannian symmetric spaces.

15.5   Symmetric spaces in general

An important class of symmetric spaces generalizing the Riemannian symmetric spaces are **pseudo-Riemannian symmetric spaces**, in which the Riemannian metric is replaced by a pseudo-Riemannian metric (nondegenerate instead of positive definite on each tangent space). In particular, **Lorentzian symmetric spaces**, i.e., n dimensional pseudo-Riemannian symmetric spaces of signature $(n-1,1)$, are important in general relativity, the most notable examples being Minkowski space, De Sitter space and anti-de Sitter space (with zero, positive and negative curvature respectively). De Sitter space of dimension n may be identified with the 1-sheeted hyperboloid in a Minkowski space of dimension $n+1$.

Symmetric and locally symmetric spaces in general can be regarded as affine symmetric spaces. If $M = G/H$ is a symmetric space, then Nomizu showed that there is a G-invariant torsion-free affine connection on M whose curvature is parallel. Conversely a manifold with such a connection is locally symmetric (i.e., its universal cover is a symmetric space). Such manifolds can also be described as those affine manifolds whose geodesic symmetries are all globally defined affine diffeomorphisms, generalizing the Riemannian and pseudo-Riemannian case.

15.5.1 Classification results

The classification of Riemannian symmetric spaces does not extend readily to the general case for the simple reason that there is no general splitting of a symmetric space into a product of irreducibles. Here a symmetric space G/H with Lie algebra

$$\mathfrak{g} = \mathfrak{h} \oplus \mathfrak{m}$$

is said to be irreducible if $\mathfrak{m}$ is an irreducible representation of $\mathfrak{h}$. Since $\mathfrak{h}$ is not semisimple (or even reductive) in general, it can have indecomposable representations which are not irreducible.

However, the irreducible symmetric spaces can be classified. As shown by Katsumi Nomizu, there is a dichotomy: an irreducible symmetric space G/H is either flat (i.e., an affine space) or $\mathfrak{g}$ is semisimple. This is the analogue of the Riemannian dichotomy between Euclidean spaces and those of compact or noncompact type, and it motivated M. Berger to classify semisimple symmetric spaces (i.e., those with $\mathfrak{g}$ semisimple) and determine which of these are irreducible. The latter question is more subtle than in the Riemannian case: even if $\mathfrak{g}$ is simple, G/H might not be irreducible.

As in the Riemannian case there are semisimple symmetric spaces with $G = H \times H$. Any semisimple symmetric space is a product of symmetric spaces of this form with symmetric spaces such that $\mathfrak{g}$ is simple. It remains to describe the latter case. For this, one needs to classify involutions σ of a (real) simple Lie algebra $\mathfrak{g}$. If $\mathfrak{g}^c$ is not simple, then $\mathfrak{g}$ is a complex simple Lie algebra, and the corresponding symmetric spaces have the form G/H, where H is a real form of G: these are the analogues of the Riemannian symmetric spaces G/K with G a complex simple Lie group, and K a maximal compact subgroup.

Thus we may assume $\mathfrak{g}^c$ is simple. The real subalgebra $\mathfrak{g}$ may be viewed as the fixed point set of a complex antilinear involution τ of $\mathfrak{g}^c$, while σ extends to a complex antilinear involution of $\mathfrak{g}^c$ commuting with τ and hence also a complex linear involution $\sigma \circ \tau$.

The classification therefore reduces to the classification of commuting pairs of antilinear involutions of a complex Lie algebra. The composite $\sigma \circ \tau$ determines a complex symmetric space, while τ determines a real form. From this it is easy to construct tables of symmetric spaces for any given $\mathfrak{g}^c$, and furthermore, there is an obvious duality given by exchanging σ and τ. This extends the compact/non-compact duality from the Riemannian case, where either σ or τ is a Cartan involution, i.e., its fixed point set is a maximal compact subalgebra.

15.5.2 Tables

The following table indexes the real symmetric spaces by complex symmetric spaces and real forms, for each classical and exceptional complex simple Lie group.

For exceptional simple Lie groups, the Riemannian case is included explicitly below, by allowing σ to be the identity involution (indicated by a dash). In the above tables this is implicitly covered by the case $kl=0$.

15.6 Weakly symmetric Riemannian spaces

Main article: Weakly symmetric space

In the 1950s Atle Selberg extended Cartan's definition of symmetric space to that of **weakly symmetric Riemannian space**, or in current terminology **weakly symmetric space**. These are defined as Riemannian manifolds M with a transitive connected Lie group of isometries G and an isometry σ normalising G such that given x, y in M there is an isometry s in G such that $sx = \sigma y$ and $sy = \sigma x$. (Selberg's assumption that s^2 should be an element of G was later shown to be unnecessary by Ernest Vinberg.) Selberg proved that weakly symmetric spaces give rise to Gelfand pairs, so that in particular the unitary representation of G on $L^2(M)$ is multiplicity free.

Selberg's definition can also be phrased equivalently in terms of a generalization of geodesic symmetry. It is required that for every point x in M and tangent vector X at x, there is an isometry s of M, depending on x and X, such that

- s fixes x;

- the derivative of s at x sends X to $-X$.

When s is independent of X, M is a symmetric space.

An account of weakly symmetric spaces and their classification by Akhiezer and Vinberg, based on the classification of periodic automorphisms of complex semisimple Lie algebras, is given in Wolf (2007).

15.7   Applications and special cases

15.7.1   Symmetric spaces and holonomy

Main article: Holonomy group

If the identity component of the holonomy group of a Riemannian manifold at a point acts irreducibly on the tangent space, then either the manifold is a locally Riemannian symmetric space, or it is in one of 7 families.

15.7.2   Hermitian symmetric spaces

Main article: Hermitian symmetric space

A Riemannian symmetric space which is additionally equipped with a parallel complex structure compatible with the Riemannian metric is called a Hermitian symmetric space. Some examples are complex vector spaces and complex projective spaces, both with their usual Riemannian metric, and the complex unit balls with suitable metrics so that they become complete and Riemannian symmetric.

An irreducible symmetric space G/K is Hermitian if and only if K contains a central circle. A quarter turn by this circle acts as multiplication by i on the tangent space at the identity coset. Thus the Hermitian symmetric spaces are easily read off of the classification. In both the compact and the non-compact cases it turns out that there are four infinite series, namely AIII, BDI with $p=2$, DIII and CI, and two exceptional spaces, namely EIII and EVII. The non-compact Hermitian symmetric spaces can be realized as bounded symmetric domains in complex vector spaces.

15.7.3   Quaternion-Kähler symmetric spaces

Main article: Quaternion-Kähler symmetric space

A Riemannian symmetric space which is additionally equipped with a parallel subbundle of $\text{End}(TM)$ isomorphic to the imaginary quaternions at each point, and compatible with the Riemannian metric, is called Quaternion-Kähler symmetric space.

An irreducible symmetric space G/K is quaternion-Kähler if and only if isotropy representation of K contains an $\text{Sp}(1)$ summand acting like the unit quaternions on a quaternionic vector space. Thus the quaternion-Kähler symmetric spaces are easily read off from the classification. In both the compact and the non-compact cases it turns out that there is exactly one for each complex simple Lie group, namely AI with $p = 2$ or $q = 2$ (these are isomorphic), BDI with $p = 4$ or $q = 4$, CII with $p = 1$ or $q = 1$, EII, EVI, EIX, FI and G.

15.7.4 Bott periodicity theorem

Main article: Bott periodicity theorem

In the Bott periodicity theorem, the loop spaces of the stable orthogonal group can be interpreted as reductive symmetric spaces.

15.8 See also

- Orthogonal symmetric Lie algebra

- Relative root system

- Satake diagram

15.9 References

- Akhiezer, D. N.; Vinberg, E. B. (1999), "Weakly symmetric spaces and spherical varieties", *Transf. Groups* **4**: 3–24, doi:10.1007/BF01236659

- van den Ban, E. P.; Flensted-Jensen, M.; Schlichtkrull, H. (1997), *Harmonic analysis on semisimple symmetric spaces: A survey of some general results*, in Representation Theory and Automorphic Forms: Instructional Conference, International Centre for Mathematical Sciences, March 1996, Edinburgh, Scotland, American Mathematical Society, ISBN 978-0-8218-0609-8

- Berger, Marcel (1957), "Les espaces symmétriques noncompacts", *Annales Scienti fiques de l'École* **74**: 85–177

- Besse, Arthur Lancelot (1987), *Einstein Manifolds*, Springer-Verlag, ISBN 0-387-15279-2 Contains a compact introduction and lots of tables.

- Borel, Armand (2001), *Essays in the History of Lie Groups and Algebraic Groups*, American Mathematical Society, ISBN 0-8218-0288-7

- Cartan, Élie (1926), "Sur une classe remarquable d'espaces de Riemann, I", *Bulletin de la Société Mathématique de France* **54**: 214–216

- Cartan, Élie (1927), "Sur une classe remarquable d'espaces de Riemann, II", *Bulletin de la Société Mathématique de France* **55**: 114–134

- Flensted-Jensen, Mogens (1986), *Analysis on Non-Riemannian Symmetric Spaces*, CBMS Regional Conference, Americal Mathematical Society, ISBN 978-0-8218-0711-8

- Helgason, Sigurdur (1978), *Differential geometry, Lie groups and symmetric spaces*, Academic Press, ISBN 0-12-338460-5 The standard book on Riemannian symmetric spaces.

- Helgason, Sigurdur (1984), *Groups and Geometric Analysis: Integral Geometry, Invariant Differential Operators, and Spherical Functions*, Academic Press, ISBN 0-12-338301-3

- Huang, Yongdong; Leung, Naichung Conan (2010). "A uniform description of compact symmetric spaces as Grassmannians using the magic square" (PDF). *Mathematische Annalen* **350** (1): 79–106. doi:10.1007/s00208-010-0549-8.

- Kobayashi, Shoshichi; Nomizu, Katsumi (1996), *Foundations of Differential Geometry, Volume II*, Wiley Classics Library edition, ISBN 0-471-15732-5 Chapter XI contains a good introduction to Riemannian symmetric spaces.

- Loos, Ottmar (1969), *Symmetric spaces I: General Theory*, Benjamin

- Loos, Ottmar (1969), *Symmetric spaces II: Compact Spaces and Classification*, Benjamin

- Nomizu, K. (1954), "Invariant affine connections on homogeneous spaces", *Amer. J. Math.* **76** (1): 33–65, doi:10.2307/2372398, JSTOR 2372398

- Selberg, Atle (1956), "Harmonic analysis and discontinuous groups in weakly symmetric riemannian spaces, with applications to Dirichlet series", *J. Indian Math. Society* **20**: 47–87

- Wolf, Joseph A. (1999), *Spaces of constant curvature* (5th ed.), McGraw–Hill

- Wolf, Joseph A. (2007), *Harmonic Analysis on Commutative Spaces*, American Mathematical Society, ISBN 0-8218-4289-7

15.10 Text and image sources, contributors, and licenses
15.10.1 Text

- **Particle physics and representation theory** *Source:* https://en.wikipedia.org/wiki/Particle_physics_and_representation_theory?oldid= 679238238*Contributors:* Michael Hardy, Charles Matthews, 4pq1injbok, Longhair, Keenan Pepper, RJFJR, Lionelbrits, Ian Pitchford, Alec.brady, MichaelSlone, Sbyrnes321, SmackBot, Bluebot, Colonies Chris, BWDuncan, Akriasas, Myasuda, Mbell, Headbomb, Shomroni, Cuzkatzimhut, Nd-brian1, Geometry guy, YohanN7, Mr. Stradivarius, 1ForTheMoney, Yobot, Gabriele Nunzio Tornetta, Niout, Citation bot, Omnipaedista,John of Reading, Brad7777, Cjean42, CsDix, 2 Hertz, Ryanexler and Anonymous: 19
- **Group theory** *Source:* https://en.wikipedia.org/wiki/Group_theory?oldid=692845828 *Contributors:* AxelBoldt, Zundark, The Anome, KF, Cwitty, Edward, Michael Hardy, Wshun, Dcljr, TakuyaMurata, Ellywa, JWSchmidt, Bogdangiusca, Poor Yorick, Rossami, Jordi Burguet Castell, Charles Matthews, Lfh, Dysprosia, Jitse Niesen, Hyacinth, Fibonacci, Phys, Bevo, Kwantus, Finlay McWalter, PuzzletChung, Grom-lakh, Romanm, Mayooranathan, Gandalf61, MathMartin, Rursus, Papadopc, ComplexZeta, Tobias Bergemann, Giftlite, Graeme Bartlett, Recentchanges, Dratman, Doshell, LiDaobing, Alberto da Calvairate~enwiki, Karl-Henner, Rich Farmbrough, FT2, Luqui, ArnoldReinhold, H00kwurm, Paul August, Tompw, Jaimedv, Adan, Obradovic Goran, Friviere, Ranveig, Masv~enwiki, HenryLi, Oleg Alexandrov, Tbsmith, Archie Paulson, OdedSchramm, Kmg90, PeterPearson, DaveApter, V8rik, BD2412, Chun-hian, Josh Parris, Rjwilmsi, Dennis Estenson II, Salix alba, Ligulem, R.e.b., Brighterorange, FlaBot, Chris Pressey, Mathbot, Margosbot~enwiki, Rune.welsh, MTC, Chobot, YurikBot, Hairy Dude, Hillman, Michael Slone, Grubber, Cate, Merlincooper, Petter Strandmark, DYLAN LENNON~enwiki, Crasshopper, Googl, Tiger-shrike, Willtron, GrinBot~enwiki, RonnieBrown, Palapa, SmackBot, Reedy, Melchoir, Scullin, Natebarney, Cessator, BiT, GBL, Bluebot, Pieter Kuiper, MalafayaBot, Ligulembot, Pilotguy, Davipo, Christopherodonovan, Lambiam, Richard L. Peterson, Utopianheaven, Mike Fikes, Tawkerbot2, Chetvorno, CRGreathouse, Ale jrb, Gregbard, Rifleman 82, Tyskis, Mungomba, Headbomb, WVhybrid, Nadav1, RobHar, NER-IUM, Escarbot, Seaphoto, M cuffa, VictorAnyakin, JAnDbot, The Transhumanist, Bongwarrior, Jakob.scholbach, CountingPine, Baccyak4H, Gabriel Kielland, David Eppstein, MaEr, David Callan, J.delanoy, Cmbankester, Indeed123, Gombang, Treisijs, Useight, Lemonaftertaste, VolkovBot, JohnBlackburne, EchoBravo, Philip Trueman, Eakirkman, Magmi, Eubulides, ArzelaAscoli, Arcfrk, Andreas Carter, Godder-sUK, Peter Stalin, Drschawrz, SieBot, Ivan Štambuk, WereSpielChequers, Viskonsas, Messagetolove, Lightmouse, JackSchmidt, NobillyT, StaticGull, Alpha Beta Epsilon, Justin W Smith, Alksentrs, Padicgroup, Bhuna71, Mspraveen, Avouac, Watchduck, Edwinconnell, Xylthixlm, Hans Adler, Vegetator, Johnuniq, TimothyRias, XLinkBot, JinJian, CàlculIntegral, Addbot, Manuel Trujillo Berges, SpellingBot, Fluffernutter, Kristine8~enwiki, Favonian, Tide rolls, Luckas-bot, Yobot, TaBOT-zerem, Julia W, Eamonster, AnomieBOT, DemocraticLuntz, Rubinbot, Μυρμηγκάκι, WinoWeritas, Citation bot, Calcio33, Auclairde, FrescoBot, Lothar von Richthofen, Orhanghazi, Sławomir Biały, Citation bot 1, Boulaur, Hard Sin, Hamtechperson, Ngyikp, D stankov, Jauhienij, Debator of mathematics, Lightlowemon, Orenburg1, FoxBot, Yger, SomeRandomPerson23, EmausBot, Fly by Night, Tommy2010, Shishir332, D15724C710N, Quondum, Kranix, Adgjdghjdety, Gottlob Gödel, ClueBot NG, Lord Roem, Ciro.santilli, HMSSolent, BG19bot, Ijgt, CimanyD, Meclee, Brad7777, Jochen Burghardt, Brirush, CsDix, Lax-fan1977, Chetan bagora, Edmundthe, KasparBot and Anonymous: 141
- **Lie group** *Source:* https://en.wikipedia.org/wiki/Lie_group?oldid=684585844 *Contributors:* AxelBoldt, Zundark, Josh Grosse, XJaM, MiguelStevertigo, Xavic69, Michael Hardy, TakuyaMurata, GTBacchus, Looxix~enwiki, Barak~enwiki, Charles Matthews, Dysprosia, Jitse Niesen,Zoicon5, David Shay, Itai, Phys, Josh Cherry, Saaska, Tobias Bergemann, Weialawaga~enwiki, Tosha, Giftlite, JamesMLane, BenFrantzDale,Lethe, Fropuff, Wgmccallum, Jason Quinn, Bobblewik, DefLog~enwiki, Lockeownzj00, Beland , Pmanderson, Abdull, Dablaze, MuDavid,Paul August, ChrisJ, Bender235, Tompw, Rgdboer, Kwamikagami, Shanes, Cherlin, Msh 210, PAR, Alex Varghese, Oleg Alexandrov, Zntrip,Joriki, Linas, Dzordzm, Isnow, SDC, AnmaFinotera, Frankie1969, Graham87, Porcher, Rjwilmsi, NatusRoma, MarSch, Salix alba, Hap-pyCamper, R.e.b., VKokielov, BMF81, Masnevets, Chobot, Algebraist, Wavelength, Hillman, RussBot, Michael Slone, KSmrq, Archelon,Buster79, Arkapravo, Smaines, Orthografer, Ekeb, Kier07, Pred, RodVance, JDspeeder1, SmackBot, Incnis Mrsi, Tom Lougheed, FlashSheri-dan, Davewild, Mhss, Kmarinas86, Bluebot, Badger014, Silly rabbit, DHN-bot~enwiki, Bears16, Akriasas, KeithB, Lambiam, Ninte, Siva1979,John, Ulner, Jim.belk, Michael Kinyon, Inquisitus, Mathchem271828, Rschwieb, Krasnoludek, Yggdrasil014, CRGreathouse, CBM, Logi-cal2u, Myasuda, Kupirijo, MotherFunctor, Dr.enh, Xantharius, Thijs!bot, Headbomb, JustAGal, RichardVeryard, RobHar, Salgueiro~enwiki,Dougher, Len Raymond, JAnDbot, Deflective, Unifey~enwiki, Homeworlds, Magioladitis, Bongwarrior, Cmelby, WhatamIdoing, Sullivan.t.j,David Eppstein, The Real Marauder, Benjamin.friedrich, David J Wilson, Jesper Carlstrom, Maproom, TomyDuby, Rocket71048576, Pidara,Fylwind, Dorftrottel, Lseixas, Borat fan, Cuzkatzimhut, Trevorgoodchild, JohnBlackburne, Ndbrian1, James.r.a.gray, Hesam7, Geometry guy,Jmath666, Eubulides, Brian Huff man, Genuine0legend, Drorata, Arcfrk, Smylei, Oscarbaltazar, YohanN7, JackSchmidt, S2000magician,Beastinwith, Mr. Stradivarius, Deciwill, Sidiropo, Leontios, Heckledpie, Cacadril, SchreiberBike, Marc van Leeuwen, MystBot, Addbot,Topology Expert, LaaknorBot, Ozob, Tanath, Tide rolls, Luckas-bot, Yobot, Ht686rg90, Niout, Amirobot, AnomieBOT, Citation bot, Arthur-Bot, Br77rino, Kaoru Itou, FrescoBot, Anterior1, Sławomir Biały, RedBot, Tinfoilcat, EmausBot, KbReZiE 12, Darkfight, Slawekb, Suslindis-ambiguator, Maschen, Zueignung, Anita5192, ClueBot NG, Mgvongoeden, Kasirbot, Helpful Pixie Bot, Daviddwd, BG19bot, CitationCleaner-Bot, Fraisière, NotWith, MathKnight-at-TAU, Suhagja, Brirush, CsDix, Sol1, Blackbombchu, Pwm86, Abitslow, Cbartondock, Victoryhuy,KasparBot, Egdunne, Referencing and Anonymous: 111
- **Classical group** *Source:* https://en.wikipedia.org/wiki/Classical_group?oldid=669636131 *Contributors:* Michael Hardy, TakuyaMurata, Charles Matthews, Semorrison, Pt, Gareth McCaughan, R.e.b., SmackBot, Incnis Mrsi, Kmarinas86, Chris the speller, Nbarth, Krasnoludek, Head-bomb, RobHar, Guy Macon, Magioladitis, David Eppstein, Leyo, Cuzkatzimhut, LokiClock, Kyle the bot, Arcfrk, YohanN7, Niceguyedc, Alexbot, J.Gowers, Addbot, Yobot, AnomieBOT, DrilBot, Slawekb, Quondum, Helpful Pixie Bot, DurwardMcDonell, CsDix, JymD, PanDTV and Anonymous: 8
- **Simple Lie group** *Source:* https://en.wikipedia.org/wiki/Simple_Lie_group?oldid=689926421 *Contributors:* Zundark, Nonenmac, Michael Hardy, TakuyaMurata, Charles Matthews, Phys, Giftlite, Fropuff, Cambyses, Sigfpe, Tomruen, MIT Trekkie, Oleg Alexandrov, GregorB, Rjwilmsi, Salix alba, R.e.b., Buster79, Kier07, SmackBot, Bluebot, Nbarth, Jim.belk, MOBle, CRGreathouse, Myasuda, Secular mind~enwiki, Thijs!bot, Escarbot, Ludvikus, R'n'B, TomyDuby, Gill110951, Red Act, Michael H 34, Arcfrk, Shadrack-dva, Addbot, Discrepancy, Niout, AnomieBOT, Omnipaedista, Erik9bot, GoingBatty, Helpful Pixie Bot, CsDix and Anonymous: 19
- **Table of Lie groups** *Source:* https://en.wikipedia.org/wiki/Table_of_Lie_groups?oldid=686760737 *Contributors:* Zundark, TakuyaMurata, Arpingstone, Fropuff, Almit39, Rich Farmbrough, Paul August, Rgdboer, John Vandenberg, R. S. Shaw, Pschemp, Oleg Alexandrov, Linas,

Salix alba, R.e.b., Mathbot, Bgwhite, KSmrq, Molinagaray, Nbarth, Vanished User 0001, Jim.belk, Sebastian Klein, Syrcatbot, Mets501, RobHar, David Eppstein, R'n'B, Pomte, Cbigorgne, Mr. Granger, Yasmar, Addbot, Matěj Grabovský, Luckas-bot, Niout, Ildeguz, Dieterich~enwiki, CsDix, Master Lenman and Anonymous: 9

- **Lie algebra** *Source:* https://en.wikipedia.org/wiki/Lie_algebra?oldid=692771994 *Contributors:* AxelBoldt, Zundark, Miguel~enwiki, Michael Hardy, Wshun, Joel Koerwer, TakuyaMurata, Suisui, Kragen, Rossami, Iorsh, Loren Rosen, Charles Matthews, Dysprosia, Michael Larsen, Grendelkhan, Phys, Tobias Bergemann, David Gerard, Weialawaga~enwiki, Tosha, Giftlite, BenFrantzDale, Lethe, Fropuff, Curps, Jeremy Henty, Jason Quinn, Python eggs, Chameleon, DefLog~enwiki, CryptoDerk, CSTAR, Pyrop, Guanabot, Pj.de.bruin, Vsmith, Gauge, Pt, Kwamikagami, Wood Thrush, Reinyday, Foobaz, Msh210, Arthena, Spangineer, Dirac1933, Drbreznjev, Oleg Alexandrov, Linas, Isnow, BD2412, NatusRoma, MarSch, Mathbot, Margosbot~enwiki, RexNL, Masnevets, YurikBot, Wavelength, Hairy Dude, Michael Slone, Lenthe, Stephenb, Grubber, Trovatore, Asimy, Crasshopper, Curpsbot-unicodify, Sbyrnes321, SmackBot, Incnis Mrsi, Grokmoo, Kmarinas86, Bluebot, Silly rabbit, Nbarth, Thomas Bliem, Chlewbot, BlackFingolfin, Noegenesis, Rschwieb, AlainD, Harold f, CmdrObot, Shirulashem, Headbomb, Second Quantization, Dachande, RobHar, B-80, Jrw@pobox.com, Deflective, Englebert, Vanish2, R'n'B, Bogey97, Maurice Carbonaro, Supermanifold, Policron, Fylwind, Cuzkatzimhut, VolkovBot, JohnBlackburne, LokiClock, Ndbrian1, Hesam7, Geometry guy, Drorata, Arcfrk, StevenJohnston, YohanN7, SieBot, Stca74, Jenny Lam, Paolo.dL, JackSchmidt, Mr. Stradivarius, Fatchat, Veromies, JP.Martin-Flatin, Count Truthstein, Addbot, Roentgenium111, Lightbot, Legobot, Luckas-bot, Yobot, Niout, Jason Recliner, Esq., DutchCanadian, Delilahblue, AnomieBOT, Twri, SassoBot, Kaoru Itou, D'ohBot, Darij, Juniuswikiae, Prtmrz, Rausch, Jkock, Adam cohenus, TobeBot, Lotje, Doctor Zook, Slawekb, Quondum, Mikhail Ryazanov, ClueBot NG, Dd314, BG19bot, Teika kazura, Walterpfeifer, Pfeiferwalter, IkamusumeFan, Flbsimas, Deltahedron, Saung Tadashi, Mark L MacDonald, Enyokoyama, CsDix, 314Username, Forgetfulfunctor00, CaptainLama, KasparBot, Texnico, Douga137 and Anonymous: 93

- **Semisimple Lie algebra** *Source:* https://en.wikipedia.org/wiki/Semisimple_Lie_algebra?oldid=677940317 *Contributors:* TakuyaMurata, CharlesMatthews, Giftlite, Fropuff, VivaEmilyDavies, Rjwilmsi, R.e.b., Masnevets, YurikBot, Wavelength, Grafen, SmackBot, Nbarth, TenPound-Hammer, Mathsci, Yggdrasil014, RobHar, David Eppstein, R'n'B, VolkovBot, JackSchmidt, Mr. Stradivarius, ELLinng, Addbot, Yobot,Niout, Omnipaedista, Sławomir Biały, Night Jaguar, ClueBot NG, Echsecutor, 𝗟𝗟𝗟𝗟, Foursided Triangle, The Disambiguator, Swbeck andAnonymous: 20

- **Homogeneous space** *Source:* https://en.wikipedia.org/wiki/Homogeneous_space?oldid=660895840 *Contributors:* Michael Hardy, Takuya-Murata, Charles Matthews, Dcoetzee, Dysprosia, Phys, Choni, Tobias Bergemann, Giftlite, Fropuff, Fleminra, Tomruen, Paul August, Gauge, Killing Vector, Oleg Alexandrov, Joriki, Linas, MarSch, Mathbot, Chobot, Eienmaru, Siddhant, YurikBot, Archelon, Silly rabbit, Nbarth, YK Times, Apon, Ixionid, Lantonov, Squids and Chips, Trigamma, YoungFrog, LokiClock, TXiKiBoT, Mr. Stradivarius, Alexbot, Nilradical, SilvonenBot, Addbot, Topology Expert, Fluffernutter, Point-set topologist, Jschnur, Fly by Night, Dewritech, Quondum, D.Lazard, Helpful Pixie Bot, Brad7777, Qetuth, Brirush, Vskrin and Anonymous: 25

- **Representation theory** *Source:* https://en.wikipedia.org/wiki/Representation_theory?oldid=687602140 *Contributors:* Zundark, TakuyaMurata, Tobias Bergemann, Unfree, Giftlite, BenFrantzDale, Frau Holle, Linas, BD2412, Wavelength, RussBot, Anomalocaris, Pred, SmackBot, Melchoir, Colonies Chris, John, Cesium 133, CmdrObot, CBM, Myasuda, RobHar, David Eppstein, R'n'B, Policron, JohnBlackburne, PaulTanenbaum, Geometry guy, Falcon8765, YohanN7, Hugh16, KathrynLybarger, Cyfal, Mr. Stradivarius, The Thing That Should Not Be, Mild Bill Hiccup, Rhubbarb, Alexbot, Addbot, Lightbot, Legobot, Andresswift, Rubinbot, Citation bot, Xqbot, Citation bot 1, Darij, I dream of horses, Kiefer.Wolfowitz, Trappist the monk, EmausBot, ZéroBot, Quondum, Helpful Pixie Bot, Beaumont877, Nosuchforever, AdventurousSquirrel, Brad7777, Majesty of Knowledge, Enyokoyama, Paritto, Jochen Burghardt, CsDix, Hamoudafg, Sol1, Liz, Melcous and Anonymous: 35

- **Representation theory of the Lorentz group** *Source:* https://en.wikipedia.org/wiki/Representation_theory_of_the_Lorentz_group?oldid= 692409821 *Contributors:* Michael Hardy, Dominus, TakuyaMurata, Charles Matthews, Giftlite, Lethe, Rgdboer, EmilJ, Jheald, Linas, Rjwilmsi, R.e.b., Bgwhite, Wavelength, Hillman, Dtrebbien, That Guy, From That Show!, SmackBot, Incnis Mrsi, Sammy1339, Mathsci, Dan Gluck, Goens, Myasuda, Mbell, Headbomb, RobHar, Magioladitis, Hullaballoo Wolfowitz, Bakken, Tumeda, David Eppstein, Schmassmann, Arturj, Leyo, Natsirtguy, M-le-mot-dit, KylieTastic, Dextercioby, Cuzkatzimhut, JohnBlackburne, Geometry guy, YohanN7, ArdClose, Mild Bill Hiccup, SchreiberBike, Count Truthstein, Yobot, Niout, AnomieBOT, Citation bot, LilHelpa, NSH002, Citation bot 1, Jonesey95, RjwilmsiBot, John of Reading, GoingBatty, Slawekb, Midas02, Quondum, Maschen, Clearlyfakeusername, Snotbot, Bibcode Bot, BG19bot, Brad7777, Toni 001, ChrisGualtieri, Stefan.Groote, CsDix, Ruby Murray, Anrnusna, Ethically Yours, Ggf4t, Ryanexler and Anonymous: 21

- **Representation theory of the Poincaré group** *Source:* https://en.wikipedia.org/wiki/Representation_theory_of_the_Poincar%C3% A9_group?oldid=602344223 *Contributors:* Charles Matthews, Phys, Oleg Alexandrov, Linas, Hailey C. Shannon, Hillman, Curpsbot-unicodify, Smack-Bot, Akriasas, Phuzion, Lmblackjack 21, Unionhawk, David Eppstein, Wing gundam, Yobot, Niout, Maschen, Razor-ockham, CsDix andAnonymous: 7

- **Representation theory of the Galilean group** *Source:* https://en.wikipedia.org/wiki/Representation_theory_of_the_Galilean_group?oldid= 690204798 *Contributors:* Michael Hardy, Charles Matthews, Phys, Jag123, Keenan Pepper, Joriki, Woohookitty, Linas, Ryan Reich, BD2412, Rjwilmsi, Wavelength, SmackBot, Kcordina, BWDuncan, Akriasas, Phuzion, CmdrObot, RobHar, Bakken, David Eppstein, R'n'B, Cuzkatzimhut, Pjoef, Wing gundam, Yobot, LilHelpa, Erik9bot, Night Jaguar, Wcherowi, Bibcode Bot, Brad7777, Fylbecatulous, CsDix and Anonymous: 8

- **List of simple Lie groups** *Source:* https://en.wikipedia.org/wiki/List_of_simple_Lie_groups?oldid=678291759 *Contributors:* Zundark, Michael Hardy, Charles Matthews, Phys, Giftlite, Fropuff, D6, ZeroOne, Oleg Alexandrov, Linas, Juan Marquez, R.e.b., Mathbot, Wavelength, Kier 07,SmackBot, Kmarinas 86, Nbarth, Ulner, Syrcatbot, N.Nahber, Nono 64, Maproom, Wesino, Nilradical, Niout, Vhsatheeshkumar, Quondum,BTotaro, CsDix and Anonymous: 5

- **Symmetric space** *Source:* https://en.wikipedia.org/wiki/Symmetric_space?oldid=678358174 *Contributors:* Michael Hardy, TakuyaMurata, Charles Matthews, Altenmann, Giftlite, LeYaYa, Fropuff, Pjacobi, Rgdboer, Teorth, Drbreznjev, Woohookitty, R.e.b., Mathbot, Masnevets, Welsh, DYLAN LENNON~enwiki, Arthur Rubin, SmackBot, Chris the speller, Nbarth, Colonies Chris, Sebastian Klein, Mathsci, CBM, MarcelBdt, RobHar, LokiClock, Nilradical, Addbot, Kilom691, Henry Godric, Point-set topologist, Sławomir Biały, Citation bot 1, Dreaming-InRed~enwiki, RjwilmsiBot, Chricho, Helpful Pixie Bot, CitationCleanerBot, Dexbot and Anonymous: 16

15.10.2 Images

- **File:Caesar3.svg** *Source:* https://upload.wikimedia.org/wikipedia/commons/2/2b/Caesar3.svg *License:* Public domain *Contributors:* Own work *Original artist:* Cepheus
- **File:Cayley_graph_of_F2.svg** *Source:* https://upload.wikimedia.org/wikipedia/commons/d/d2/Cayley_graph_of_F2.svg *License:* Public domain *Contributors:* ? *Original artist:* ?
- **File:Circle_as_Lie_group.svg** *Source:* https://upload.wikimedia.org/wikipedia/commons/8/82/Circle_as_Lie_group.svg *License:* Public domain *Contributors:* self-made with en:Inkscape *Original artist:* Oleg Alexandrov
- **File:Commutative_diagram_SO(3, _1)_latex.svg** *Source:* https://upload.wikimedia.org/wikipedia/en/a/a1/Commutative_diagram_SO%2C_1%29_latex.svg *License:* CC-BY-SA-3.0 *Contributors:*
Using LaTeX + pdfcrop + pdf2svg
Original artist:
Slawekb
- **File:Connected_Dynkin_Diagrams.svg** *Source:* https://upload.wikimedia.org/wikipedia/commons/3/3b/Connected_Dynkin_Diagrams.svg *License:* GFDL *Contributors:* Created by me by copying Image:ConnectedDynkinDiagrams.png *Original artist:* R. A. Nonenmacher
- **File:Cyclic_group.svg** *Source:* https://upload.wikimedia.org/wikipedia/commons/5/5f/Cyclic_group.svg *License:* CC BY-SA 3.0 *Contributors:*
- Cyclic_group.png *Original artist:*
- derivative work: Pbroks13 (talk)
- **File:Dirac_4.jpg** *Source:* https://upload.wikimedia.org/wikipedia/commons/c/cf/Dirac_4.jpg *License:* Public domain *Contributors:* http://nobelprize.org/nobel_prizes/physics/laureates/1933/dirac.html *Original artist:* Nobel Foundation
- **File:Einstein_en_Lorentz.jpg** *Source:* https://upload.wikimedia.org/wikipedia/commons/c/c6/Einstein_en_Lorentz.jpg *License:* Public domain *Contributors:* ? *Original artist:* ?
- **File:ExponentialMap-01.png** *Source:* https://upload.wikimedia.org/wikipedia/commons/0/06/ExponentialMap-01.png *License:* Public domain *Contributors:* ? *Original artist:* ?
- **File:Fifths.png** *Source:* https://upload.wikimedia.org/wikipedia/commons/c/ce/Fifths.png *License:* CC-BY-SA-3.0 *Contributors:* ? *Original artist:* ?
- **File:Finite_Dynkin_diagrams.svg** *Source:* https://upload.wikimedia.org/wikipedia/commons/0/0c/Finite_Dynkin_diagrams.svg *License:* CC BY-SA 3.0 *Contributors:* Created by me by copying File:Connected_Dynkin_Diagrams.svg *Original artist:* Tomruen
- **File:Hermann_Weyl_ETH- Bib_Portr_00890.jpg** *Source:* https://upload.wikimedia.org/wikipedia/commons/7/78/Hermann_Weyl_ETH-Portr_00890.jpg *License:* CC BY-SA 3.0 *Contributors:* ETH-Bibliothek Zürich, Bildarchiv *Original artist:* ETH Zürich
- **File:IM_Gelfand.jpg** *Source:* https://upload.wikimedia.org/wikipedia/en/5/5e/IM_Gelfand.jpg *License:* Fair use *Contributors:* ? *Original artist:* ?
- **File:Liealgebra.png** *Source:* https://upload.wikimedia.org/wikipedia/commons/d/d2/Liealgebra.png *License:* Public domain *Contributors:* http://en.wikipedia.org/wiki/File:Liealgebra.png *Original artist:* Phys
- **File:Lorentz_group_commutative_diagram_2.svg** *Source:* https://upload.wikimedia.org/wikipedia/commons/8/83/Lorentz_group_comm diagram_2.svg *License:* CC0 *Contributors:* Own work *Original artist:* Maschen
- **File:Mergefrom.svg** *Source:* https://upload.wikimedia.org/wikipedia/commons/0/0f/Mergefrom.svg *License:* Public domain *Contributors:* ? *Original artist:* ?
- **File:Miri2.jpg** *Source:* https://upload.wikimedia.org/wikipedia/commons/d/db/Miri2.jpg *License:* CC BY-SA 3.0 *Contributors:* Own work *Original artist:* J:136401
- **File:Question_book-new.svg** *Source:* https://upload.wikimedia.org/wikipedia/en/9/99/Question_book-new.svg *License:* Cc-by-sa-3.0 *Contributors:*
Created from scratch in Adobe Illustrator. Based on Image:Question book.png created by User:Equazcion *Original artist:*
Tkgd2007
- **File:Richard_Brauer.jpg** *Source:* https://upload.wikimedia.org/wikipedia/commons/7/78/Richard_Brauer.jpg *License:* CC BY-SA 2.0 de *Contributors:* http://owpdb.mfo.de/detail?photo_id=467 *Original artist:* Jacobs, Konrad
- **File:Root_system_A1xA1.svg** *Source:* https://upload.wikimedia.org/wikipedia/commons/2/2c/Root_system_A1xA1.svg *License:* CC-BY-SA-3.0 *Contributors:*
- Root-system-A1xA1.png *Original artist:* Root-system-A1xA1.png: User:Maksim
- **File:Rubik'{}s_cube.svg** *Source:* https://upload.wikimedia.org/wikipedia/commons/a/a6/Rubik%27s_cube.svg *License:* CC-BY-SA-3.0 *Contributors:* Based on Image:Rubiks cube.jpg *Original artist:* This image was created by me, Booyabazooka
- **File:Sophus_Lie.jpg** *Source:* https://upload.wikimedia.org/wikipedia/commons/a/ab/Sophus_Lie.jpg *License:* Public domain *Contributors:* ? *Original artist:* Ludwik Szacinski
- **File:Standard_Model_charges.svg** *Source:* https://upload.wikimedia.org/wikipedia/commons/f/f1/Standard_Model_charges.svg *License:* CC BY-SA 3.0 *Contributors:* Own work, Created from Garret Lisi's Elementary Particle Explorer *Original artist:* Cjean42
- **File:Torus.png** *Source:* https://upload.wikimedia.org/wikipedia/commons/1/17/Torus.png *License:* Public domain *Contributors:* ? *Original artist:* ?

- **File:Wigner.jpg** *Source:* https://upload.wikimedia.org/wikipedia/commons/e/ef/Wigner.jpg *License:* Public domain *Contributors:* http://nobelprize.org/nobel_prizes/physics/laureates/1963/wigner-bio.html *Original artist:* Nobel foundation

- **File:Wilhelm_Karl_Joseph_Killing.jpeg** *Source:* https://upload.wikimedia.org/wikipedia/commons/5/57/Wilhelm_Karl_Joseph_Killing.jpeg *License:* Public domain *Contributors:* http://www- history.mcs.st- andrews.ac.uk/Mathematicians/Killing.html *Original artist:* Unknown<ahref='//www.wikidata.org/wiki/Q4233718' title='wikidata:Q4233718'><img alt='wikidata:Q4233718' src='https://upload.wikimedia.org/wikipedia/commons/thumb/f/ff/Wikidata- logo.svg/20px- Wikidata- logo.svg.png' width='20' height='11' srcset='https://upload.wikimedia.org/wikipedia/commons/thumb/f/ff/Wikidata- logo.svg/30px- Wikidata- logo.svg.png 1.5x, https://upload.wikimedia.org/wikipedia/commons/thumb/f/ff/Wikidata- logo.svg/40px- Wikidata- logo.svg.png 2x' data-file-width='1050' data-file-height='590' /></a>

15.10.3 Content license

www.ingramcontent.com/pod-product-compliance
Lightning Source LLC
Chambersburg PA
CBHW080658190526
45169CB00006B/2165